Day-by-Day Math Thinking Routines in Fourth Grade

Day-by-Day Math Thinking Routines in Fourth Grade helps you provide students with a review of the foundational ideas in math, every day of the week! Based on the bestselling *Daily Math Thinking Routines in Action*, the book follows the simple premise that frequent, rigorous, engaging practice leads to mastery and retention of concepts, ideas, and skills. These worksheet-free, academically rigorous routines and prompts follow the grade level priority standards and include whole group, individual, and partner work. The book can be used with any math program, or for small groups, workstations, or homework.

Inside you will find:

- 40 weeks of practice
- 1 activity a day
- 200 activities total
- Answer Key

For each week, the Anchor Routines cover these key areas: Monday: General Thinking Routines; Tuesday: Vocabulary; Wednesday: Place Value; Thursday: Fluency; and Friday: Problem Solving. Get your students' math muscles moving with the easy-to-follow routines in this book!

Dr. Nicki Newton has been an educator for over 30 years, working both nationally and internationally with students of all ages. She has worked on developing Math Workshop and Guided Math Institutes around the country; visit her website at www.drnickinewton.com. She is also an avid blogger (www.guidedmath.wordpress.com), tweeter (@drnickimath) and Pinterest pinner (www.pinterest.com/drnicki7).

Also Available from Dr. Nicki Newton

(www.routledge.com/K-12)

Day-by-Day Math Thinking Routines in Kindergarten: 40 Weeks of Quick Prompts and Activities

Day-by-Day Math Thinking Routines in First Grade: 40 Weeks of Quick Prompts and Activities

Day-by-Day Math Thinking Routines in Second Grade: 40 Weeks of Quick Prompts and Activities

Day-by-Day Math Thinking Routines in Third Grade: 40 Weeks of Quick Prompts and Activities

Day-by-Day Math Thinking Routines in Fifth Grade: 40 Weeks of Quick Prompts and Activities

Daily Math Thinking Routines in Action: Distributed Practices Across the Year

Leveling Math Workstations in Grades K–2: Strategies for Differentiated Practice

Fluency Doesn't Just Happen with Addition and Subtraction: Strategies and Models for Teaching the Basic Facts
With Ann Elise Record and Alison J. Mello

Mathematizing Your School: Creating a Culture for Math Success
Co-authored by Janet Nuzzie

Math Problem Solving in Action: Getting Students to Love Word Problems, Grades K-2

Math Problem Solving in Action: Getting Students to Love Word Problems, Grades 3–5

Guided Math in Action: Building Each Student's Mathematical Proficiency with Small-Group Instruction

Math Workshop in Action: Strategies for Grades K-5

Math Running Records in Action: A Framework for Assessing Basic Fact Fluency in Grades K-5

Math Workstations in Action: Powerful Possibilities for Engaged Learning in Grades 3–5

Day-by-Day Math Thinking Routines in Fourth Grade

40 Weeks of Quick Prompts and Activities

Dr. Nicki Newton

NEW YORK AND LONDON

First published 2021
by Routledge
52 Vanderbilt Avenue, New York, NY 10017

and by Routledge
2 Park Square, Milton Park, Abingdon, Oxon, OX14 4RN

Routledge is an imprint of the Taylor & Francis Group, an informa business

Library of Congress Cataloging-in-Publication Data
Names: Newton, Nicki, author.
Title: Day-by-day math thinking routines in fourth grade: 40 weeks of quick prompts and activities / Dr. Nicki Newton.
Description: New York : Routledge, 2020. |
Identifiers: LCCN 2019051701 (print) | LCCN 2019051702 (ebook) | ISBN 9780367901745 (hardback) | ISBN 9780367901707 (paperback) | ISBN 9781003022923 (ebook)
Subjects: LCSH: Mathematics—Study and teaching (Elementary)—Activity programs.
Classification: LCC QA135.6 .N48459 2020 (print) | LCC QA135.6 (ebook) | DDC 372.7/049—dc23
LC record available at https://lccn.loc.gov/2019051701
LC ebook record available at https://lccn.loc.gov/2019051702

ISBN: 978-0-367-90174-5 (hbk)
ISBN: 978-0-367-90170-7 (pbk)
ISBN: 978-1-003-02292-3 (ebk)

Typeset in Palatino
by Swales & Willis, Exeter, Devon, UK

Contents

Meet the Author

Dr. Nicki Newton has been an educator for over 30 years, working both nationally and internationally, with students of all ages. Having spent the first part of her career as a literacy and social studies specialist, she built on those frameworks to inform her math work. She believes that math is intricately intertwined with reading, writing, listening and speaking. She has worked on developing Math Workshop and Guided Math Institutes around the country. Most recently, she has been helping districts and schools nationwide to integrate their State Standards for Mathematics and think deeply about how to teach these within a Math Workshop Model. Dr. Nicki works with teachers, coaches and administrators to make math come alive by considering the powerful impact of building a community of mathematicians who make meaning of real math together. When students do real math, they learn it. They own it, they understand it, and they can do it. Every one of them. Dr. Nicki is also an avid blogger (www.guidedmath. wordpress. com) and Pinterest pinner (https://www.pinterest.com/drnicki7/).

Introduction

Welcome to this exciting new series of daily math thinking routines. I have been doing thinking routines for years. People ask me all the time if I have these written down somewhere. So, I wrote a book. Now, that has turned into a grade level series so that people can do them with prompts that address their grade level standards. This is the anti-worksheet workbook!

The goal is to get students reflecting on their thinking and communicating their mathematical thinking with partners and the whole class about the math they are learning. Marzano (2007)[1] notes that:

> initial understanding, albeit a good one, does not suffice for learning that is aimed at long-term retention and use of knowledge. Rather, students must have opportunities to practice new skills and deepen their understanding of new information. Without this type of extended processing, knowledge that students initially understand might fade and be lost over time.

The daily math thinking routines in this book focus on taking Depth of Knowledge activity level 1 activities, to DOK level 2 and 3 activities (Webb, 2002)[2]. Many of the questions are open. For example, we turn the traditional elapsed time question on its head. Instead of asking students "Mark left his house at 3:15 p.m. and he came back 20 minutes later. When did he come back?" Inspired by Marion Smalls (2009)[3] we ask, "An activity takes 20 minutes. When could it have started and when could it have ended?"

In this series, we work mainly work on priority standards, although we do address some of the supporting and additional standards. This book is not intended to cover every standard. Rather it is meant to provide ongoing daily review of the foundational ideas in math. There is a focus for each day of the week.

- Monday: General Thinking Routines
- Tuesday: Vocabulary
- Wednesday: Place Value
- Thursday: Fluency (American and British Number Talks Number Strings)
- Friday: Problem Solving

On Monday the focus is on general daily thinking routines (What Doesn't Belong?, True or False?, Convince Me), that review various priority standards from the different domains (Geometry, Algebraic Thinking, Counting, Measurement, Number Sense). Every Tuesday there is an emphasis on Vocabulary because math is a language and if you don't know the words then you can't speak it. There is a continuous review of foundational words through different games (Tic Tac Toe, Match, Bingo), because students need at least 6 encounters with a word to own it. On Wednesday there is often an emphasis on Place Value. Thursday is always some sort of Fluency routine (American or British Number Talks and Number Strings). Finally, Fridays are Problem Solving routines.

The book starts with a review of third grade priority standards and then as the weeks progress the current grade level standards are integrated throughout. There is a heavy emphasis on

1 Marzano, R. J. (2007). *The art and science of teaching: A comprehensive framework for effective instruction*. ASCD: Virginia.

2 Webb, N. (March 28, 2002) "Depth-of-Knowledge Levels for four content areas," unpublished paper.

3 Small, M. (2009). *Good questions: Great ways to differentiate mathematics instruction*. Teachers College Press: New York.

multidigit operations, fractions and decimals. There is also an emphasis on geometry concepts and some data and measurement. There are various opportunities to work with word problems throughout the year.

Throughout the book there is an emphasis on the mathematical practices/processes (SMP, 2010[4]; NCTM, 2000[5]). Students are expected to problem solve in different ways. They are expected to reason by contextualizing and decontextualizing numbers. They are expected to communicate their thinking to partners and the whole group using the precise mathematical vocabulary. Part of this is reading the work of others, listening to others' explanations, writing about their work and then speaking about their work and the work of others in respectful ways. Students are expected to model their thinking with tools and templates. Students are continuously asked to think about the pattern and structure of numbers as they work through the activities.

These activities focus on building mathematical proficiency as defined by the NAP 2001[6]. These activities focus on conceptual understanding, procedural fluency, adaptive reasoning, strategic competence and a student's mathematical disposition. This book can be used with any math program. These are jump starters to the day. They are getting the math muscle moving at the beginning of the day.

Math routines are a form of "guided practice." Marzano notes that although the:

> guided practice is the place where students—working alone, with other students, or with the teacher—engage in the cognitive processing activities of organizing, reviewing, rehearsing, summarizing, comparing, and contrasting. However, it is important that all students engage in these activities. Rosenshine, (cited p.7 in Marzano, 2007)

These are engaging, standards-based, academically rigorous activities that provide meaningful routines that develop mathematical proficiency. The work can also be used for practice within small groups, workstations and also sent home home as questions for homework.

We have focused on coherence from grade to grade, rigor of thinking, and focus on understanding and being able to explain the math the students are doing. We have intended to take deeper dives into the math, not rushing to the topics of the next grade but going deeper into the topics of the grade at hand (see Figures 1.1–1.4). Here is our criteria for selecting the routines:

- Engaging
- Easy to learn
- Repeatable
- Open-ended
- Easy to differentiate (adapt and extend for different levels).

4 The Standards of Mathematical Practice. "Common Core State Standards for Mathematical Practice." Washington, D.C.: National Governors Association Center for Best Practices, Council of Chief State School Officers, 2010. Retrieved on December 1, 2019 from: www.corestandards.org/Math/Practice.

5 National Council of Teachers of Mathematics. (2000). *Principles and standards for school mathematics*. Reston, VA: National Council of Teachers of Mathematics.

6 Kilpatrick, J., Swafford, J., and Findell, B. (eds.) (2001). *Adding it up: Helping children learn mathematics*. Washington, DC: National Academy Press.

Figure 1.1 Talking about the Routine!

Monday: General Thinking Routines

3 x ___ = 12

Jen said that the answer is 36. Kelly said the answer is 4. Who do you agree with? Why? How would you change the equation to make it so that you can agree with the person you disagreed with?

Tuesday: What Doesn't Belong?

Which one doesn't belong?

rhombus	rectangle
trapezoid	square

Wednesday: Guess My Number

I am a 2 digit number.
I am more than 12.
I am less than 20.
I am not odd.
I am a multiple of 7
Who am I?

Thursday: Number Talk

Multiply 2 numbers that have a product near 150.

Friday: Problem Solving

Mary bought a rug with an area of 24 square feet. What could the perimeter have been?

Figure 1.2 The Math Routine Cycle of Engagement

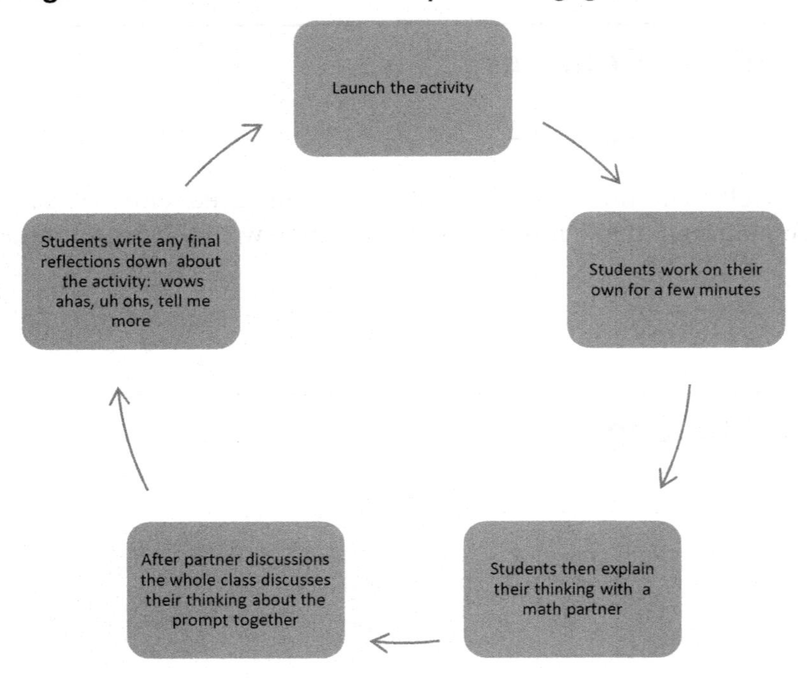

Step 1: Students are given the launch prompt. The teacher explains the prompt and makes sure that everyone understands what they are working on.

Step 2: They are given a few minutes to work on that prompt by themselves.

Step 3: The next step is for students to work with a math partner. As they work with this partner, students are expected to listen to what their partner did as well as explain their own work.

Step 4: Students come back together as a whole group and discuss the math. They are encouraged to talk about how they solved it and the similarities and differences between their thinking and their partner's thinking.

Step 5: Students reflect on the prompt of the day, thinking about what Wowed them, what made them say "Ah-ha", what made them say "Uh-oh", what made them say, "I need to know more about this."

Thinking Activities

These are carefully planned practice activities to get students to think. They are **not meant to be used as a workbook**. This is a thinking activity book. The emphasis is on students doing their own work, explaining what they did with a partner and the sharing out to the entire class.

Overview of the Routines

Monday Routines – General Thinking Routines (Algebraic Thinking, Measurement, Data, Geometry)

- Always, Sometimes, Never
- Break It Up!
- Convince Me!
- Draw That!
- Guess My Number
- It Is/It Isn't
- Input/Output Table
- Legs and Feet
- Magic Square
- Missing Number
- Number Line It!
- Open Array Puzzle
- Pattern/Skip Counting
- Reasoning Matrices
- Subtraction Puzzle
- 3 Truths and a Fib
- True or False?
- 2 Arguments
- Venn Diagram
- What Doesn't Belong?
- Why Is It Not?

Tuesday Routines – Vocabulary

- Vocabulary Bingo
- Convince Me!
- Frayer Model
- It Is/It Isn't
- 1 Minute Essay
- Talk and Draw
- Vocabulary Tic Tac Toe
- Venn Diagram
- Vocabulary Brainstorm
- Vocabulary Match
- What Doesn't Belong?

Wednesday Routines – Place Value/Fractions

- 3 Truths and a Fib
- Bingo
- Greater Than, Less Than, in Between
- Convince Me!
- Find and Fix the Error
- Fraction of the Day
- Guess My Number
- How Many More to

- Model It
- Money Combinations
- Number Bond It!
- Number Line It!
- Number of the Day
- Pattern/Skip Counting
- Place Value Puzzle
- Rounding
- Start At… Get To… by
- Venn Diagram
- What Doesn't Belong?

Thursday Routines – Number Talk

- British Number Talk
- Find and Fix the Error
- Open Array Puzzle
- Number Talk
- Number Strings
- Number Talk Puzzle
- What's Missing?

Friday Routines – Problem Solving

- Equation Match
- Fill in the Problem!
- Make Your Own Problem!
- Model It!
- Picture That!
- Time Problem
- Regular Word Problem
- Sort That!
- What's the Question? (3 Read Protocol)
- What's the Story? Here's the Model
- What's the Story? Here's the Graph

Figure 1.3 Overview of the Routines

Routine	Purpose	Description
Always, Sometimes, Never	This routine focuses on students reasoning about whether a statement is always, sometimes or never true.	In this routine, students are given a statement and they have to argue and prove their thinking about if the statement is always, sometimes or never true.
Break It Up!	In this routine, students work on the distributive property.	In this routine, students are given an expression and they have to sketch it and then break it apart using the distributive property.
British Number Talk	This routine focuses on students thinking about their thinking.	Students have to choose their own problems and discuss how they are going to solve them. They must name the way they did it, either in their head, with a model or with an algorithm.
Convince Me!	This routine focuses on students reasoning about different topics. They have to convince their peers about specific statements.	Students are given different things to think about like statements or equations and they have to convince their peers that they are correct.
Draw That!	This routine focuses on students drawing about the topic.	Students are given a topic and then asked to draw pictures that describe that topic.
Equation Match	This routine focuses on students thinking about which operation they would use to solve a problem. It requires that they reason about the actions that are happening in the problem and then what they are required to do to solve the problem.	Students are trying to pair the word problem and the equation.
Fill in the Problem!	In this routine, students have to fill in numbers and make up and solve their own word problem.	Students fill in the blanks with numbers that they choose and then model and solve the word problem.

Routine	Purpose	Description
Find and Fix the Error	This routine requires that students analyze the work of others and discuss what went well or what went wrong. The purpose of the routine is not only to get students to identify common errors but also to get them to justify their own thinking about the problem.	Students think about a problem that may be correct or incorrect either by themselves, with a partner or with the whole group. They have to figure out why it is done incorrectly or correctly and discuss.
Fraction of the Day	This routine focuses on students thinking about and modeling fractions.	Students are given a fraction and they have to write it in word form, draw a visual model, plot it on a number line and discuss it in relationship to other fractions.
Frayer Model	This routine is meant to get students talking about concepts. They are supposed to talk about the definition, what happens in real life, sketch an example and also give nonexamples.	Students are given a template with labels. They work through the template writing and drawing about the specified topic.
Greater Than, Less Than, in Between	In this routine, students specifically talk about numbers in terms of greater than, less than and in between each other.	Students are thinking about the number relationships and filling in boxes based on those relationships.
Guess My Number	This routine gives students a variety of clues about a number and asks the students to guess which number it might be, given all the clues. Students have to use their understanding of place value and math vocabulary to figure out which number is being discussed.	Students are given various clues about a number and they must use the clues to guess which number it is.
How Many More to	In this routine, students are asked to tell how many more to a specific number. Again, this is another place value routine, asking students to reason about numbers on the number line.	Students are given a specified number and they have to tell how many more to that number.
Input/Output Table	This routine focuses on getting students to think about patterns.	Students have to fill in the number of the input/output table and in some cases create their own from scratch.

Routine	Purpose	Description
It Is/It Isn't	This routine can be used in a variety of ways. Students have to look at the topic and decide what it is and what it isn't. It is another way of looking at example, nonexample.	Students discuss what something is and what it isn't.
Legs and Feet	Legs and Feet is a great arithmetic routine which gets students to use various operations to figure out how many animals there could be.	Students look at different animals and think about how many legs and feet there could be given that number of animals.
Magic Square	In this fluency routine, students are working with math puzzles to figure out missing numbers.	There are a few different ways to do magic squares. One way is for students to figure out what the magic number is going horizontally, vertically and diagonally.
Make Your Own Problem	In this routine, students have to write their own problems.	Students have to write problems based on an equation or based on a topic.
Missing Number	This routine focuses on students thinking about missing numbers.	Many of the missing number activities require that students reason about what number it should be.
Model It	In this word problem routine, students are focusing on representing word problems in a variety of ways.	Students have to represent their thinking about a word problem with various models.
Money Combinations	This routine focuses on students knowledge of money.	The money routines have students counting and comparing money quantities.
Number Bond It!	In this routine, students are working on decomposing numbers in a variety of ways.	Students use number bonds to break apart numbers in different ways.
Number of the Day	This routine focuses on students representing and modeling numbers in a variety of ways.	This activity has a given number and students have to represent that number in different ways.
Number Line It	This activity focuses on sequencing numbers correctly.	Students have to put numbers in the correct sequence on the number path.

Routine	Purpose	Description
Number Talk	This activity focuses on number sense. Students compose and decompose numbers as well as add and subtract numbers. Students discuss solving different problems in different ways.	There are a few different ways that students do this activity. One of the ways is the teacher works with the students on solving a problem in a variety of ways. Another activity is that the teacher gives the students number strings around a specific concept for example subtracting 1 from a number and students work those problems and discuss the strategy.
Number Talk Puzzle	In this routine, students reason about numbers.	Students have to decide which numbers are missing to complete the problem.
Number Strings	In this routine, students are looking at the relationship among a set of problems.	Students work out the different problems and think about and discuss the various strategies they are using.
Open Array Puzzle	In this routine, students think about open arrays.	Students have to decide which numbers are missing to make the open array true.
Patterns/Skip Counting	In this routine, students focus on patterns.	Many of the pattern activities require students to fill in a pattern and then make their own patterns.
Picture That!	In this routine, students discuss a picture.	Many of these activities require that students look at a picture and make up a math word problem about the picture.
Place Value Puzzle	In this routine, students reason about numbers.	Students have to decide which numbers are missing to correctly complete the problem.
Reasoning Matrix	In this routine, students have to reason about a variety of things to figure out which ones go together.	Students are asked to reason about people given specific facts.
Regular Word Problems	In this routine, students have to solve traditional word problems.	Students are expected to solve the problem, model their thinking and write the equation.
Rounding	In this routine, students work on rounding.	The rounding activities ask students to come up with the numbers that can be rounded rather than giving students a number to round.

Routine	Purpose	Description
Sort That!	In this routine, students are reasoning about what type of problem they are looking at.	Students sort the problems and decide which one is the designated type that they are looking for.
Start at … Get to … By	In this routine, students focus on patterns and skip counting.	Students practice skip counting to different numbers.
Subtraction Puzzle	In this routine, students have to reason about the numbers in the puzzle.	Students have to decide which numbers to fill in the puzzle to make it true.
Talk and Draw	In this routine, students have to talk and draw about the topic.	Students are given a topic and are expected to discuss their thinking with math sketches.
Time Problem	This routine is an open question, where students work with elapsed time.	Students have to write an elapsed time problem.
True or False?	This routine focuses on students reasoning about what is true or false.	Students are given different things to think about like statements about shapes or equations and they have to state and prove whether they are true or false.
Venn Diagram	In this routine, students fill out a Venn diagram with specific criteria.	Students are given a Venn diagram that they must fill out based on specific criteria.
Vocabulary Bingo	This is the traditional bingo game with a focus on vocabulary.	Students play vocabulary bingo but they have to discuss the vocabulary and make drawings or write definitions to show they know what the word means.
Vocabulary Brainstorm	In this routine, students have to brainstorm about vocabulary words.	Students have to write about different vocabulary words.
Vocabulary Match	In this routine, the focus is on working with the math vocabulary from across the year.	Students match the vocabulary with the definition.
Vocabulary Tic Tac Toe	In this routine, students are working on math vocabulary words from across the year.	Students play tic tac toe by taking turns choosing a square and then sketching or writing on the side to illustrate the word. Whoever gets 3 in a row first wins.

Routine	Purpose	Description
Why Is It Not?	This routine focuses on students looking at error patterns and correcting them.	Students have to look at error patterns and then discuss what the correct answer should be and prove why it should be that.
What Doesn't Belong?	This is a reasoning activity where students have to choose which objects they can group together and why. The emphasis is on justification.	Students have 4 squares. They have to figure out which object does not belong.
What's Missing?	In this routine, students have to reason about numbers in equations.	Students have to decide which number makes the equation true.
What's the Question? (3 Read Protocol)	The purpose is for students to slow down and consider all of the parts of the word problem.	Students have to read the problem 3 times. The first time the focus is on the context. The second time the focus is on the numbers. The third time they focus on asking questions that would make sense given the context.
What's the Story?	This routine focuses on students making sense of models and graphs.	Students have to look at the model and make up a story that matches it.
Why Is It Not?	In this routine, students are expected to reason about problems.	Students have to explain why the incorrect answer is not true.
1-Minute Essay	In this routine, students have to think about, discuss and write about a concept.	Students write about a topic, their friend adds information and then they write some more about the topic.
2 Arguments	In this reasoning routine, students are thinking about common errors that students make when doing various math tasks like missing numbers, working with properties and working with the equal sign.	Students listen to the way 2 different students approached a problem, decide who they agree with and defend their thinking.
3 Truths and a Fib	In this routine, students have to consider whether or not statements are true or false.	Students are given 3 statements that are true and 1 statement that is false. They have to discuss why it is false.

Questioning is the Key

To Unlock the Magic of Thinking, You Need Good Questions

Figure 1.4

Launch Questions (Before the Activity)	Process Questions (During the Activity)
♦ What is this prompt asking us to do? ♦ How will you start? ♦ What are you thinking? ♦ Explain to your math partner, your understanding of the question. ♦ What will you do to solve this problem?	♦ What will you do first? ♦ How will you organize your thinking? ♦ What might you do to get started? ♦ What is your strategy? ♦ Why did you....? ♦ Why are you doing that? ♦ Is that working? Does it make sense? ♦ Is that a reasonable answer? ♦ Can you prove it? ♦ Are you sure about that answer? ♦ How do you know you are correct?
Debrief Questions (After the Activity)	**Partner Questions (Guide Student Conversations)**
♦ What did you do? ♦ How did you get your answer? ♦ How do you know it is correct? ♦ Can you prove it? ♦ Convince me that you have the correct answer. ♦ Is there another way to think about this problem?	♦ Tell me what you did. ♦ Tell me more about your model. ♦ Tell me more about your drawing. ♦ Tell me more about your calculations. ♦ Tell me more about your thinking. ♦ Can you prove it? ♦ How do you know you are right? ♦ I understand what you did. ♦ I don't understand what you did yet.

Daily Routines

Week 1 Teacher Notes

Monday: What Doesn't Belong?

When doing this activity, have the students do the calculations (in their journals, on scratch paper or on the activity page). Then, have them share their thinking with a friend. Then, pull them back to the group. So, in this routine it is important to focus on language as well for the descriptions. The language should be something like: 5 + 3 + 4 has a sum of 12 and the other problems all equal 8.

In Set B, most of the problems are subtraction. Some students will say it's the multiplication problem. Accept that answer and then look for others. Always validate and affirm what students say. Work it into an ongoing conversation. For example: Yes that is true. What else might we think about this set? You want students to be able to say "The difference of all the other expressions is 7 but the difference between 50 and 33 is 17.

Tuesday: Vocabulary Match

This vocabulary match has many of the third grade words that students should know. Often when reviewing vocabulary it is good to review the grade level words mixed, meaning not by a specific category. Students should say the word and then find the matching definition. They should have some minutes to do this on their own and then an opportunity to go over their thinking with their math partner. Then, after about 5 minutes, come back together as a group and discuss the thinking. Ask students which words were tricky and which ones were easy. Also ask them were there any that they didn't recognize, that they have never seen before? Have them draw a little sketch by each word to help them remember the word.

Wednesday: Convince Me!

This routine is about getting students to defend and justify their thinking. Be sure to emphasize the language of reasoning. Students should focus on proving it with numbers, words and pictures. They should say things like:

This is true! I can prove it with....

This is the difference because....

I am going to use _________ to show my thinking.

I am going to defend my answer by _________.

I proved my thinking using addition.

Thursday: Number Talk

In this number talk you want the students to discuss their thinking with strategies and models. Ask students about the strategies that they might use.

Possible responses:

Add the hundreds, then tens then ones.

Give and Take (Compensation) – Take 2 from 76 and make 128 into 130. Add 74 and 9 to make 83 and then add 83 to 130 to make 213. Discuss how this makes it a much easier problem when we work with tens.

Friday: What's the Question? (3 Read Protocol)

The focus of today is to do a 3 read problem with the students. *It is important to read the problem 3 times out loud as a choral read with the students.*

First Read: (Stop and visualize! What do you see?) What is this story about? Who is in it? What are they doing?

Second Read: What are the numbers? What do they mean?

Third Read: What are some possible questions we could ask about this story?

Possible Questions:

How many marbles does he have altogether?

How many orange marbles does he have?

How many green marbles does he have?

How many more green marbles does he have than orange ones?

*Note: Focus on the vocabulary. Use different words for the sum (altogether, total).

Focus on different types of comparative language so students get comfortable with words and phrases like: How many more? How many less? How many more to get the same amount as? How many fewer?

Week 1 Activities

Monday: What Doesn't Belong?

Look at the sets of problems below. Pick the one that doesn't belong in each set.

A.

4×2	$80 \div 10$
$5 + 3 + 4$	$12 - 4$

B.

$20 - 13$	$30 - 23$
$(3 \times 5) - 8$	$50 - 33$

Tuesday: Vocabulary Match

Match the word with the definition.

sum	the answer to a multiplication problem
product	the amount left over when one number is subtracted from another number
quotient	numbers that are multiplied together to find a product
difference	the answer to a division problem
factor	the result of adding numbers or quantities together

Wednesday: Convince Me!

Prove it with numbers, words and/or pictures!

You have to defend your thinking with numbers, words and pictures. Discuss this with your math partner and then the whole group.

This is true! I can prove it with....

This is the same because....

I am going to use __________ to show my thinking.

I am going to defend my answer by __________.

A. $\frac{2}{3}$ is greater than $\frac{1}{2}$

B. $\frac{6}{8}$ is equal to $\frac{3}{4}$

Thursday: Number Talk

Discuss this with a partner and then with your class.

What are some ways to think about:

$$76 + 128 + 9$$

Friday: What's the Question? (3 Read Protocol)

Read the problem 3 times with your class. The first time talk about what the problem is about. The second time talk about what the numbers mean. The third time talk about at least 2 questions you could ask about this story. Write them down. Discuss with your classmates.

Tom has 3 boxes of orange marbles with 5 in each. He also has 3 boxes of green marbles, with 10 in each.

Week 2 Teacher Notes

Monday: Magic Square

This routine is great for practice. They are intriguing, fun and fast-paced. Students want to get the answer and they get right to work. In this Magic Square students are trying to figure out the target number (the one that the digits add up to no matter which way you calculate them – horizontally, vertically or diagonally).

Have the students work on it on their own, then share their thinking with a partner. Then, bring everyone back to the whole group and have them discuss it.

Tuesday: Vocabulary Tic Tac Toe

These are quick partner energizers. Read all the words together. Then go! Students have 7 minutes to play the game. They do rock, paper, scissors to start. They take turns choosing a word and explaining it to their partner. Then, they have to do a sketch or something to show they understand the word. Everybody should play the first game, if they have time, they can play the next one.

It is important to call everyone back together at the end and talk about the vocabulary. Briefly go over the vocabulary, this is all third-grade vocabulary.

Wednesday: Number Line It!

In this routine the students are working with fractions. This is actually a much more difficult skill than it seems. Students have trouble thinking about how to place the numbers on the number line as accurately as possible. There should be an emphasis on doing this absolutely accurately. So, students should be encouraged to think about where the numbers might go, put a tick on the line, discuss it with their partner, defend their thinking and then to go back and put the numbers on the line. After students have placed their numbers on the line, the whole class comes back together and discusses what has been done. The more that students do this, the better they get.

Thursday: British Number Talk

There are a variety of numbers in the circles so that students can choose a variety of problems. Encourage them to take risks. Talk about picking maybe some easy problems and then some tricky ones.

Friday: Make Your Own Problem!

Students have to write a word problem where the answer is 18 square feet.

Monday: Magic Square

Look at the square below. What is the magic number? It must be the same sum no matter what direction you add it in.

4	9	2
3	5	7
8	1	6

Tuesday: Vocabulary Tic Tac Toe

Play rock, paper, scissors to see who goes first. Then take turns, picking a square, saying what it means, drawing or writing something about the word on the side and then marking the word with an x or an o. Whoever gets 3 in a row first wins.

Game 1

quotient	addend	difference
perimeter	centimeter	liter
gram	area	factor

Game 2: Say the Name of the Shape

Wednesday: Number Line It!

Place these numbers on the number line! Be as exact as possible.

$\frac{1}{2}$ $\frac{2}{8}$ $\frac{4}{4}$ $\frac{1}{4}$ $\frac{5}{4}$

Compare these with your partner. Explain your thinking. How do you know you are correct?

Thursday: British Number Talk

Pick a number from each circle. Then, decide how you are going to solve it using subtraction. Write the problem under the way you solved it. For example: 218 – 118. I can do that in my head.

I can do it in my head.	I can do it with a model.	I can do it using a written strategy or algorithm.

218	470
94	150
1000	165
578	611
226	108
219	100

	118	319
	257	26
	417	14
25	30	215
19	10	50

Friday: Make Your Own Problem!

Write a word problem about area. The answer is 18 square feet.

Week 3 Teacher Notes

Monday: Always, Sometimes, Never

In this routine you are trying to get students to reason. So they should try out what they think by writing down at least 3 examples and discussing them with their partner. After they have discussed their thinking with their partner, the class should come back together and discuss their thinking further. The emphasis should be on proving whatever they believe.

Tuesday: Frayer Model

Students fill out the diagram based on the word.

Wednesday: Number of the Day

Number of the day is important and reviews the place value skills. There are both closed and open items in the routine. Give the students about 5 minutes to work on this and then discuss it with their math partner. Then, come back together as a class and talk about what students did.

Thursday: Number Strings

Students should discuss the string as a whole class. You want the emphasis to be on the relationships between the numbers. Students should be asked if they know one of the base facts and how it helps them with the other facts. Sometimes, people call these "helper" facts. Although these are first- and second grade-strategies, many fourth graders are still very shaky with these, thus the ongoing review.

Friday: Model It

The focus for this Friday is the open number line. We want students to be very comfortable with the open number line. Encourage them to use it often, along with the tape diagram. Focus on the idea of jumping tens and multiples of tens. They can do this in a variety of ways. They could add up all the tens and then the ones. They could also start at a number and then jump to the nearest ten from there. For example, 89 + 1 would get you to 90 and then add 50 and then from 140 add 60 to get to 200 and add 6 more. Also, students could just add 50, 60 and 80 to get 190 and then add 16 more.

Week 3 Activities

Monday: Always, Sometimes, Never

Read the statement. Decide if it is always, sometimes, or never true. Discuss your thinking with your math partner and the whole group.

Quadrilaterals are always parallelograms.

Tuesday: Frayer Model

Fill in the boxes based on the word.

Quadrilateral

Description	Example
Picture	Non-example

Wednesday: Number of the Day

Fill in the boxes below based on that number.

1,199

Number word	Show 2 addition sentences that make 1,199.
Show 2 subtraction sentences that make 1,199.	Is it prime or composite? Is it odd or even? Round it to the nearest 100:

Thursday: Number Strings

Discuss the patterns that you see in these problems with a math partner and then the whole group.

String 1	String 2
27 + 27	478 + 95
27 + 26	787 + 35
27 + 25	187 + 26
27 + 24	218 + 47
	18 + 78

Friday: Model It

Solve in 2 different ways.

399 + 144

A. Model this problem on the number line.

B. Decompose the numbers to find the sum.

C. Use models to justify your answer.

Week 4 Teacher Notes

Monday: Magic Square

In this routine students have to fill in the missing numbers. Remember that magic squares are made to foster flexibility. Give the students a few minutes to work on it by themselves and then have them discuss their thinking with their math partner. Then, come back together as a class and have the students discuss their thinking.

Tuesday: It Is/It Isn't

In this routine you want students to be focusing on the vocabulary. Encourage students to use the word bank. This is a scaffold only though, to get them started. The conversation might sound something like this: It is a 2-digit number. It is not a 1-digit number. It is an odd number. It is not an even number. It is greater than 20. It is less than 90. It is in between 30 and 60. It is 10 more than 43. It is 10 less than 63.

Wednesday: Greater Than, Less Than, in Between

In this routine, students are thinking about the number relationships. Have the students do it on their own and then talk with their math partners. Then, bring them back together as a class and discuss it.

Thursday: Number Talk

This is a typical number talk where students are thinking about the ways in which they can solve this subtraction problem. You want students to think about partial differences, counting up, and compensation. You want students to talk about how if you add 1 to each number you get an easier problem: 1,405 – 1,300.

Friday: Picture That!

Students look at the box of donuts and write any type of story they want to. They can work on this with their partner. Then they should be ready to share it with the whole class.

Week 4 Activities

Monday: Magic Square

Figure out which numbers will make this magic square have a total of 15 in all directions.

2		6
	5	
	3	

Tuesday: It Is/It Isn't

Discuss what quadrilaterals are and what they are not!

It Is	It Isn't

Wednesday: Greater Than, Less Than, in Between

Fill in the boxes discussing these numbers.

$\frac{1}{6}$ $\frac{1}{3}$ $\frac{6}{6}$

Name a fraction less than $\frac{1}{3}$	Name a fraction in between $\frac{1}{3}$ and $\frac{6}{6}$	Name a fraction that is less than $\frac{1}{6}$
Name a fraction greater than $\frac{1}{3}$	Name a fraction greater than $\frac{6}{6}$	Name a fraction in between $\frac{1}{6}$ and $\frac{6}{6}$

Thursday: Number Talk

What are some ways to subtract 1,299 from 1,404?

Friday: Picture That!

Tell a fraction story about this box of donuts. Write the equation that can be used to represent the story you came up with.

Story:

Equation:

Week 5 Teacher Notes

Monday: 2 Arguments

Students look at this problem and discuss it with their math partner. The emphasis should be on proving their thinking. Discuss how what they said is true. Then, everyone will come back together and discuss it.

Tuesday: Vocabulary Match

Say all the words together as a class and get students to think about if they know these words, if they are familiar, or if they are completely unfamiliar. Then, have students turn and work with their partners on the match. Come back together as a class and discuss the words.

Wednesday: Money Combinations

Money is challenging for many students. They worked on it a great deal in second grade, so this is a way of reviewing that content and building flexibility with numbers. Students should do it on their own, then share their thinking with a partner and finally come back together and discuss with the class.

Thursday: Number Talk

Give students about 5 minutes to work on their own or with partners to come up with some problems, hopefully from each category. Make sure that they stretch themselves. You don't want them to only choose the easy problem. Then, students should share what they did with the class.

Friday: Make Your Own Problem!

Give students about 5 minutes to do their own fill-in problems and share with their math partner. Their partner has to tell them if their problem makes sense. Then, some students will share their thinking with the class.

Week 5 Activities

Monday: 2 Arguments

Look at the problem. Discuss with a neighbor. Be ready to talk with the whole class.

$4 \times \_\_\_\_ = 12$

John said the answer was 48.

Maria said the answer was 3.

Who do you agree with?

Why?

What would John or Maria need to do in order to convince you to agree with them?

Tuesday: Vocabulary Match

Match the vocabulary with the example.

ten-thousands	
hexagon	
thousands	500 + 80 + 5
expanded form	14,091
quadrilateral	75,160

Wednesday: Money Combinations

Raul had $1.89. Show 3 different combinations of coins and bills he could have had.

Way 1	Way 2	Way 3

Thursday: Number Talk

You are subtracting two numbers. The numbers have a difference close to 125. What pairs of numbers might you be subtracting?

Friday: Make Your Own Problem!

Use 3-digit numbers.

Toy store A had _______ marbles in stock. Toy store B had _______ more marbles than toy store A. How many more marbles did toy store B have than toy store A?

Model it!

Number sentence (equation): ____________________

Week 6 Teacher Notes

Monday: It Is/It Isn't

In this routine you want students to be focusing on the vocabulary. It is a 5 digit number. It is not odd.

Tuesday: 1-Minute Essay

Give students the designated part of the time to write and then share and then write again and then share out with the class.

Wednesday: Find and Fix the Error

Students have to look at the problem with their math partner and discuss what is incorrect. Then, figure out how they would fix it. Be ready to come back and discuss it with the class.

Thursday: Number Strings

Students should discuss the string as a whole class. You want the emphasis to be on the relationships between the numbers. In this string we are working on what happens with 9s. Students should recognize that we can just compensate and make the 9 into a 10 and add from there. With practice, students do this naturally.

Friday: Equation Match

Students have to match the story with the correct equation.

Week 6 Activities

Monday: It Is/It Isn't

$(2 \times 10{,}000) + (3 \times 100) + (4 \times 10) + (6 \times 1)$

It Is	It Isn't

Tuesday: 1-Minute Essay

(For 30 seconds) Write everything you can about division. Use numbers, words and pictures.

(15 seconds) Now switch with a neighbor and add 1 thing to their list.

(15 seconds) Now add 1 more thing to your list.

Wednesday: Find and Fix the Error

John did this. The answer is wrong. Find and fix the error.

$$\begin{array}{r} 1000 \\ -\ 598 \\ \hline 598 \end{array}$$

1. What is wrong?

2. Why can't you do what John did?

3. Fix it.

4. Explain your thinking to your partner and then the whole group.

Thursday: Number Strings

$9 + 4$
$89 + 24$
$129 + 64$
$8{,}999 + 1{,}794$

Friday: Equation Match

Pick the equation that matches the above problem.

Lucy had 4 boxes of crayons. Each box had 12 crayons. She took the crayons and placed an equal number of crayons into 3 supply boxes. How many crayons did she put in each supply box?

A. $4 \times 12 + 3 = ?$
B. $12 \div 4 \times 3 = ?$
C. $4 \times 12 \div 3 = ?$
D. Answer not here.

Week 7 Teacher Notes

Monday: 3 Truths and a Fib

This is a measurement reasoning activity.

Tuesday: Vocabulary Brainstorm

Students have to think and then write about the word in all of the bubbles. They first share their thinking with a partner and then share their thinking with the class.

Wednesday: Pattern/Skip Counting

Students fill in the pattern, discuss their thinking with a neighbor and then discuss as a whole class. It is really important that students make their own pattern as well. There must be time given for students to do that.

Thursday: Number Talk Puzzle

Students think about what numbers they need to add to the puzzle to make the puzzle true.

Friday: Make Your Own Problem!

Students have to write a word problem where the answer is 3 with a remainder of 2. This is a reasoning problem. Lead the discussion as a whole class and reason about it. Start with problems that have 3 as an answer and then have students think about how to come up with a problem from there.

Week 7 Activities

Monday: 3 Truths and a Fib

Which one is false? Why? Explain to your neighbor and then the group.

A door is about 8 feet tall.	A baby elephant can weigh about 90 kilos.
A car is about 40 yards long.	A lizard can be about 6 inches long.

Tuesday: Vocabulary Brainstorm

In each thought cloud write or draw something that has to do with multiplication.

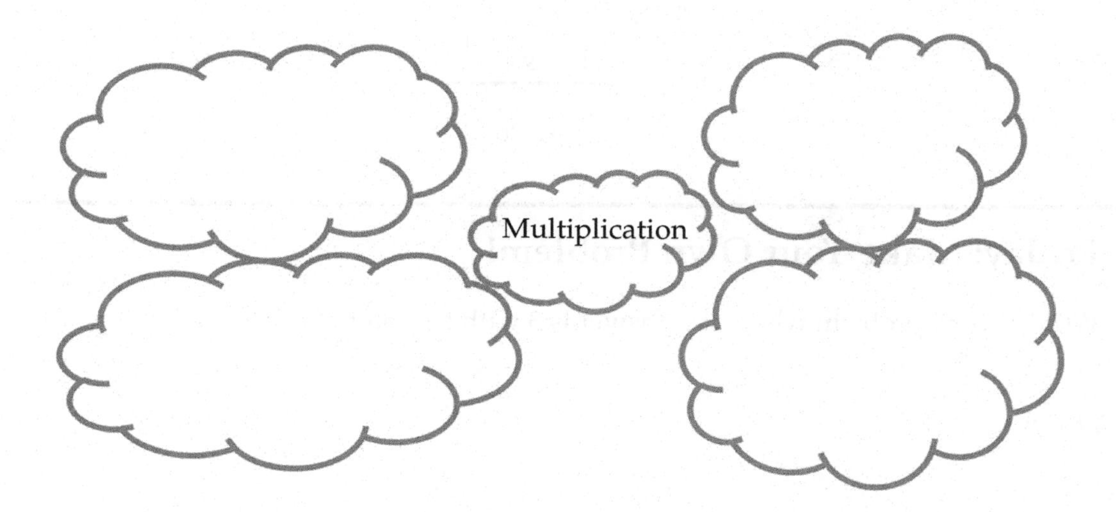

Wednesday: Pattern/Skip Counting

Complete the patterns

1. $\frac{1}{2}$, $\frac{3}{4}$, ____, $1\frac{1}{4}$, $1\frac{1}{2}$, ____, 2, ____, ____
2. $\frac{2}{3}$, $1\frac{1}{3}$, ___, ___, ___
3. Make your own:

   ____, ____, ____, ____, ____, ____, ____, ____

Thursday: Number Talk Puzzle

What are the missing numbers? Fill in numbers that make the puzzle true.

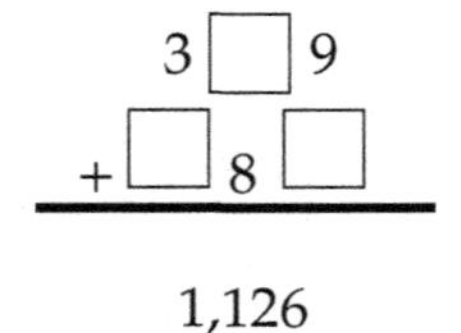

1,126

Friday: Make Your Own Problem!

Write a word problem where the answer is 3 with a remainder of 2.

Story:

Model:

Equation:

Week 8 Teacher Notes

Monday: True or False?

Many times, students are quick to dismiss this as not a hexagon, because it is unfamiliar as a hexagon. So, have students think about it and debate it with a partner before discussing it with the whole class. It is really important to follow up by having the students to draw another hexagon. Encourage them to draw an irregular hexagon.

Tuesday: Vocabulary Tic Tac Toe

These are quick partner energizers. Read all the words together. Then go! Students have 7 minutes to play the game. They do rock, paper, scissors to start. They take turns choosing a word and explaining it to their partner. Then, they have to do a sketch or something to show they understand the word. Everybody should play the first game, if they have time, they can play the next one.

It is important to call everyone back together at the end and talk about the vocabulary. Briefly go over the vocabulary, this is all third-grade vocabulary.

Wednesday: Number Bond It!

Number bonds are important because they build flexibility. Students should be thinking about how to compose and decompose numbers in a variety of ways. Have the students work on this by themselves first and then discuss with a partner and finally with the entire class.

Thursday: Number Talk

Give students about 5 minutes to work on their own or with partners to come up with some problems, hopefully from each category. Make sure that they stretch themselves. You don't want them to only choose the easy problem. Then, students should share what they did with class.

Friday: Model It

The focus here is that students use the tape diagram. Although we want students to use a variety of models, they have to learn them first so that they have a repertoire to choose from. In second grade, they started working on tape diagrams so it is good to work on these every chance you get.

Week 8 Activities

Monday: True or False?

Decide which statement is true:

All of these are hexagons.
Some of these are hexagons.
None of these are hexagons.

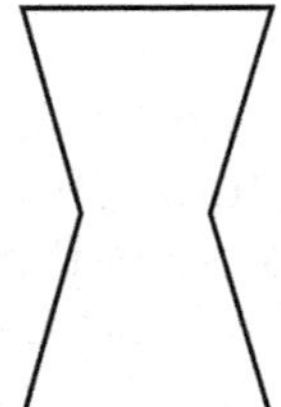 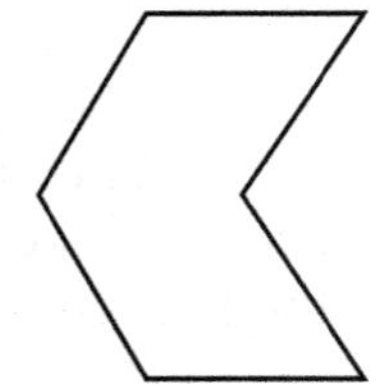 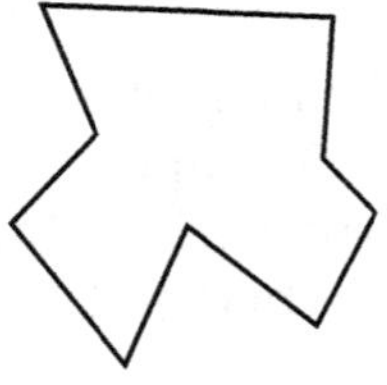 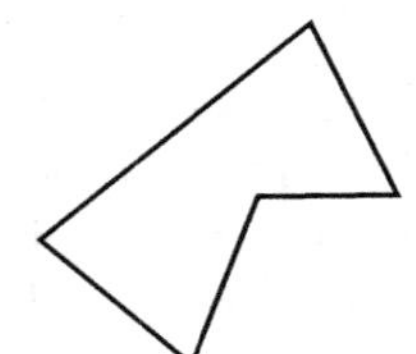 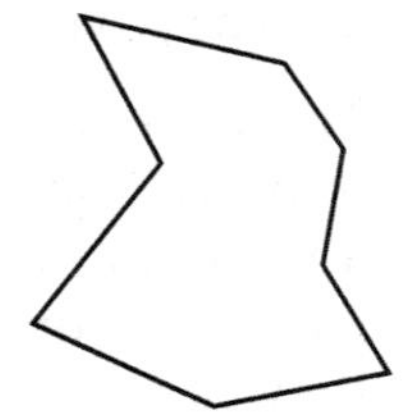

1. Think about it.
2. Share your thinking with a friend. Defend your answer.
3. Share your thinking with the group.

Tuesday: Vocabulary Tic Tac Toe

Play rock, paper, scissors to see who goes first. Then take turns, picking a square, saying what it means, drawing or writing something about the word on the side and then marking the word with an x or an o. Whoever gets 3 in a row first wins.

quotient	centimeter	hundreds
divisor	thousands	digit
equation	division	expanded form

area	perimeter	yard
feet	inch	compare
kilometer	pentagon	cylinder

Wednesday: Number Bond It!

Show how to decompose apart 3,942 in 3 different ways!

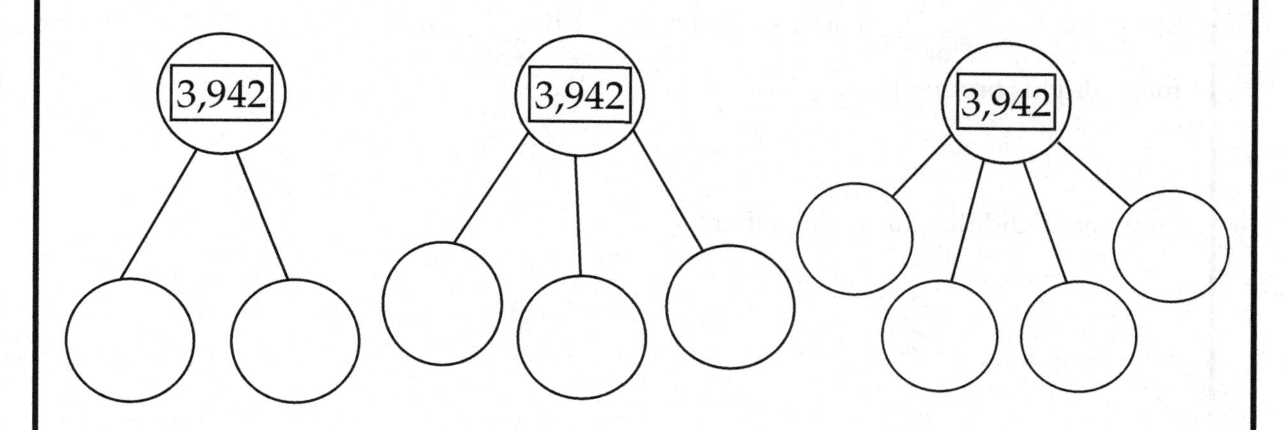

Thursday: Number Talk

Complete the following equations so that they each have a sum or differences between 555 and 602.

A. ______ – _______ = ________
B. ______ + ______ = ______
C. _____ + _____ + ______ = _______
D. ______ + ________ – ________ = ______

Friday: Model It

Choose the numbers. Solve the problem.

Story: Joe had ______ marbles. His brother had ______ times as many as he did. How many did his brother have?

(Under each blank: 2 or 3)

How many did they have altogether?

Tape diagram:

Equation:

Week 9 Teacher Notes

Monday: Convince Me!

This routine is about getting students to defend and justify their thinking. Be sure to emphasize the language of reasoning. Students should focus on proving it with numbers, words and pictures. They should say things like:

This is true! I can prove it with . . .
This is the same because . . .
I am going to use __________ to show my thinking.
I am going to defend my answer by __________.

Tuesday: Vocabulary Match

Students have to match the word with the definition.

Wednesday: Fraction of the Day

Students fill in the template based on the fraction of the day.

Thursday: Number String

Students should discuss the string as a whole class. You want the emphasis to be on the relationships between the numbers. In this number string students are thinking about what it means to subtract 7, 8 or 9. How do you adjust the number to make it an easier problem. So when we think about 142 – 27 we can add 3 to each number and make it an easier problem- 145 – 30. Practicing these strategies is so important so that they become internalized in student thinking.

Friday: Model It

The students have to model the word problem and then solve it.

Week 9 Activities

Monday: Convince Me!

Defend your thinking with numbers, words and pictures. Discuss with your partners and then the whole class.

Convince me that:
$\frac{5}{3} = \frac{2}{3} + \frac{2}{3} + \frac{1}{3}$
This is true! I can prove it with….
This is the same because….
I am going to use __________ to show my thinking.
I am going to defend my answer by __________.

Tuesday: Vocabulary Match

Match the word and the definition.

equivalent	shows the number of parts taken from the denominator
remainder	to break apart
denominator	the same as
numerator	the part that is left over of a division problem
decompose	the part that names how many equal pieces the whole has been divided into

Wednesday: Fraction of the Day

Fill in the boxes based on the fraction.

$$\frac{5}{8}$$

<table>
<tr><td>Word form.</td><td>How much more is needed to get to 1 whole?</td><td>Draw a model.</td></tr>
<tr><td colspan="2">Plot it on the number line.

Plot a fraction that is greater than this fraction. Add a fraction that is less than this fraction. Plot an equivalent fraction on the number line.</td><td>____ = ____

____ > ____

____ < ____</td></tr>
</table>

Thursday: Number String

Look for a pattern. Discuss.

12 – 9
22 – 8
142 – 27
262 – 139
373 – 248
1584 – 359

Friday: Model It

Grandma Betsy made some pies. She cut one into 4 pieces. She cut another into 8 pieces. If she wants to give the same amount from each cake, how many eighths will be equivalent to three-fourths?

Week 10 Teacher Notes

Monday: Reasoning Matrices

Reasoning matrices are great. Students should read them and work on them with a partner and then share their thinking with the whole class.

Tuesday: 1-Minute Essay

Students have to write everything they can about polygons using numbers, words and pictures. They then switch with a partner, who adds to their work and then they finish up by adding one more thing to their work.

Wednesday: Number of the Day

Number of the day is important and reviews the place value skills. There are both closed and open items in the routine. Give the students about 5 minutes to work on this and then discuss it with their math partner. Then, come back together as a class and talk about what students did.

Thursday: Number Talk

This is a typical number talk where students are thinking about the ways in which they can solve these problems. They should discuss the relationships they notice.

Friday: Make Your Own Problem!

Students have to write a word problem about milliliters where the answer is 2 liters. They should do this on their own and share it with their partner to see if their story makes sense. Then, they will discuss their thinking with the whole class.

Week 10 Activities

Monday: Reasoning Matrices

	Pepperoni	Cheese	Chicken	Mushroom	Veggies	Pepperoni and Pineapple
Jenny						
Jamal						
Miguel						
Kelly						
Maria						
Grace						

Use the clues to figure out who ate which pizza. Jenny doesn't like anything on her pizza except sauce and cheese. Grace loves vegetables. Kelly loves both fruit and meat. Miguel loves chicken. Jamal wants meat but not chicken. Maria loves vegetables but she only wants 1 on her pizza at a time.

Tuesday: 1-Minute Essay

(For 30 seconds) Write everything you can about polygons. Use numbers, words and pictures.

(15 seconds) Now switch with a neighbor and add 1 thing to their list.

(15 seconds) Now add 1 more thing to your list.

Wednesday: Number of the Day

Fill in the boxes based on the number.

7,2⑦<u>7</u>

Word form:	10 more:	10 less:
Expanded form:	_____ + _____ = 7,277	_____ – _____ = 7,277
100 more:	How many more to 10,000?	Odd or even?
1,000 more:	100 less:	7,247 – _____ = ______

What is the relationship between the value of the digits circled and underlined?

Thursday: Number Talk

What patterns do you notice? How does this help you to solve these problems?

4×10
4×20
4×30
4×40
4×70
4×80

Friday: Make Your Own Problem!

Tell a word problem about milliliters, where the answer is 2 liters.

Story:

Model:

Equation:

Week 11 Teacher Notes

Monday: Venn Diagram

Venn diagrams are great for getting students to compare and contrast. Have students work with partners and then debrief as a whole class.

Tuesday: Vocabulary Bingo

Today the focus of the vocabulary is on multiplication and division with some other words mixed in. Students have worked with all of these words in prior grades. Remember that you are trying to normalize these words in the students' vocabulary. Again, the idea is to keep math vocabulary up and present in the minds of students.

Wednesday: How Many More to

In this routine, students start at a specific number and they have to tell how many more to the target number.

Thursday: What's Missing?

Students discuss the missing number. The focus should be on how they arrived at the answer. They have to convince their classmates that they are correct.

Friday: What the Question? (3 Read Protocol)

The focus of today is to do a 3 read problem with the students. *It is important to read the problem 3 times out loud as a choral read with the students.*

First read: (Stop and visualize! What do you see?) What is this story about? Who is in it? What are they doing?

Second read: What are the numbers? What do they mean?

Third read: What are some possible questions we could ask about this story?

Possible questions:

How much did Mike and Tom eat altogether?

How much did Joe eat?

Who ate most of the candy bar?

Note: Focus on the vocabulary. Use different words for the sum (altogether, total).

Focus on different types of comparative language so students get comfortable with words and phrases like: How many more? How many less? How many more to get the same amount as? How many fewer?

Week 11 Activities

Monday: Venn Diagram

Fill in the Venn diagram.

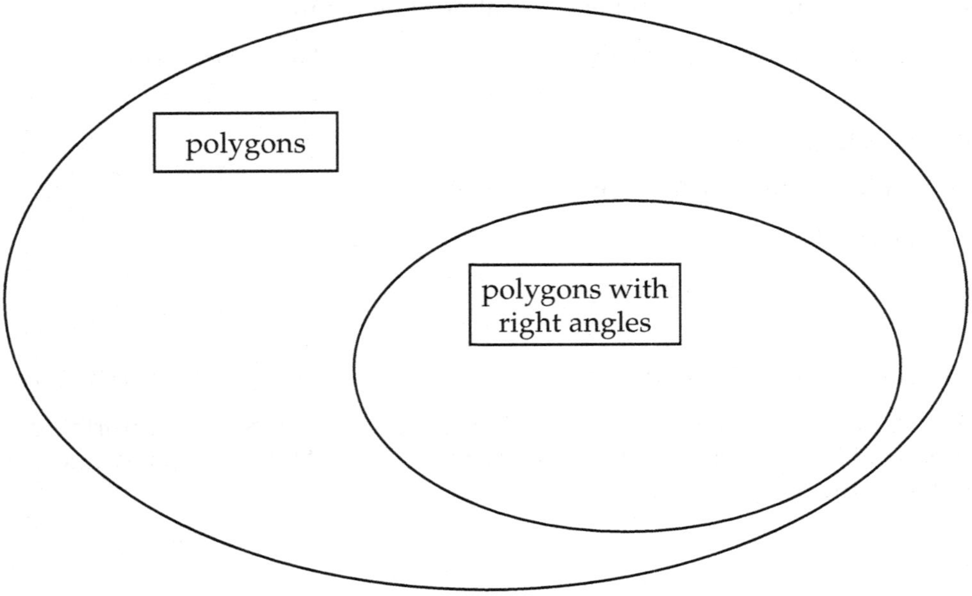

polygons

polygons with right angles

Tuesday: Vocabulary Bingo

Put these words on the bingo board in different spaces. Do not put them in the order they appear. Listen as your teacher calls out the definitions and cross out that space. Whoever gets 4 in a row horizontally, vertically or as a postage stamp (4 in a corner) first wins.

Words: remainder, quotient, sum, divisor, dividend, addition, addend, product, difference, factor, multiple, even, polygon, gram, quadrilateral, liter.

Wednesday: How Many More to

For each problem, start with the number and state how many more to the designated number.

A.	B.
1. Start with 25 get to 100 2. Start with 179 get to 200 3. Start with 589 get to 1000	1. Start with $\frac{1}{5}$ get to 1 2. Start with $\frac{2}{12}$ get to $\frac{4}{6}$ 3. Start with $\frac{2}{3}$ get to 1

Thursday: What's Missing?

A. $8 \div \_\_\_\_ = 4$

B. $\_\_\_\_ \div 10 = 100$

C. $25 \div \_\_\_\_ = 5$

D. $\_\_\_ \div \_\_\_\_ = \_\_\_\_$

Explain how to find the answer.

Friday: What's the Question? (3 Read Protocol)

Read the problem 3 times with your class. The first time talk about the story. The second time discuss the numbers. The third time think of 2 questions that you could ask about this story. Write them down. Discuss with your classmates.

Mike and his brothers shared a candy bar. Mike ate $\frac{2}{5}$ of it. Tom ate $\frac{1}{5}$ of it. Joe ate the rest.	What are some questions that you can ask?

Week 12 Teacher Notes

Monday: Legs and Feet

Students should work with a partner. They should draw pictures and work with tables.

Tuesday: Vocabulary Brainstorm

Students have to think and then write about the word in all of the bubbles. They first share their thinking with a partner and then share their thinking with the class.

Wednesday: 3 Truths and a Fib

Students think about it on their own. They then work with a partner and prepare to share their thinking with the whole class.

Thursday: Number Talk

You want students to think about counting up or adding 72 to each number to get a more friendly number.

Friday: Make Your Own Problem!

Students have to write a word problem about division with a remainder. They should do this on their own and share it with their partner to see if their story makes sense. Then, they will discuss their thinking with the whole class.

Week 12 Activities

Monday: Legs and Feet

There is a chicken and a cow. Answer the questions, based on your reasoning about the animals. Do math sketches or write down numbers if you need to.

A. How many legs?	B. If there are 12 legs and there has to be a chicken and a cow, how many animals and what type are there?
C. If there are 14 legs and there has to be a chicken and a cow, how many animals and what type are there?	D. If there are 20 legs and there has to be a chicken and a cow, how many animals could there be?
E. What if we added a cricket? How many legs now? What if we had 24 legs, what is a possible combination?	F. If we had 30 legs, what could be a possible combination of animals?

Tuesday: Vocabulary Brainstorm

In each thought cloud write or draw something that has to do with measurement.

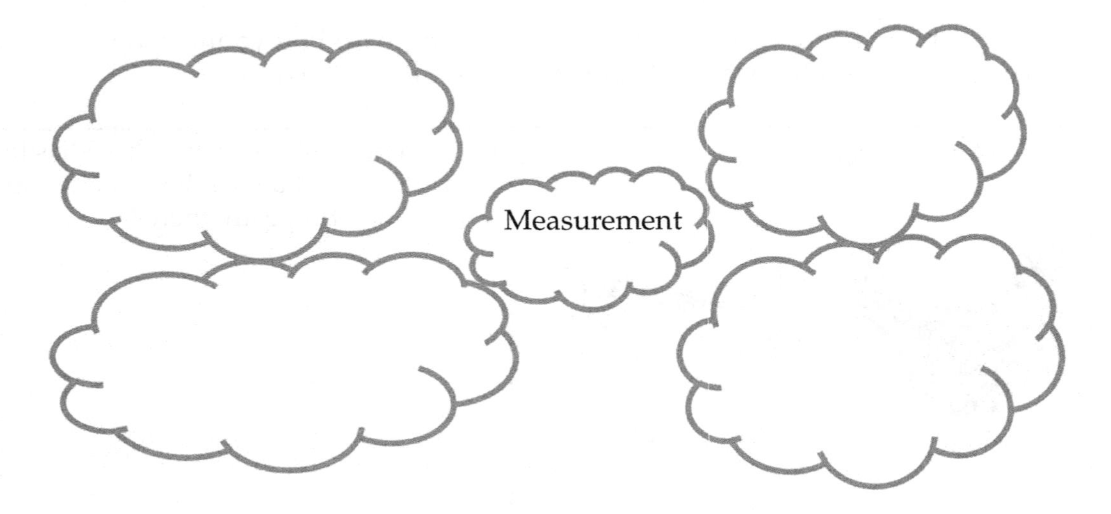

Wednesday: 3 Truths and a Fib

Look at the statements below. Decide which one is false. Discuss with a friend.

$\frac{1}{2} = \frac{5}{10}$	$\frac{2}{4} = \frac{5}{8}$
$\frac{3}{4} = \frac{6}{8}$	$\frac{2}{3} = \frac{4}{6}$

Thursday: Number Talk

What are some ways to subtract 10,000 – 1,828?

Friday: Make Your Own Problem!

Write a division word problem with a remainder.

Story:

Model:

Equation:

Week 13 Teacher Notes

Monday: Patterns/Skip Counting

Students have to figure out the pattern and then make their own.

Tuesday: What Doesn't Belong?

Students have to figure out which words do not belong in the sets of words.

Wednesday: Rounding

These are open problems where students are expected to come up with numbers that round to a particular number. This is a level above just giving a student a number and having them round that number. Here they have to know more about rounding.

Thursday: Number Talk

We want students to talk about what this expression means. It's 5 groups of 7. If you haven't gotten to formal multiplication yet, have the students think about this problem and how to solve it using repeated addition. Think about the turn around fact of this problem. Also, if you are there in the curriculum, discuss how we could solve this using the distributive problem.

Friday: What's the Question? (3 Read Protocol)

The focus of today is to do a 3 read problem with the students. *It is important to read the problem 3 times outloud as a choral read with the students.*

First read: (Stop and visualize! What do you see?) What is this story about? Who is in it? What are they doing?

Second read: What are the numbers? What do they mean?

Third read: What are some possible questions we could ask about this story?

Possible questions:

How many gold rings did they have?

How many silver rings did they have?

How many gold and silver rings did they have altogether?

Note: Focus on the vocabulary. Use different words for the sum (altogether, total).

Focus on different types of comparative language so students get comfortable with words and phrases like: how many more, how many less, how many more to get the same amount as, how many fewer?

Week 13 Activities

Monday: Pattern/Skip Counting

Jamal wrote a pattern. His rule was to add 1 to a number and then multiply it by 2. Which pattern below follows his rule?

Part A.

A. 4, 10, 13, 115.

B. 5, 12, 27, 135.

C. 9, 20, 42, 86.

D. Not here.

Part B.
Make your own pattern that follows Jamal's rule.

Tuesday: What Doesn't Belong?

Look at each set. Decide which word does not belong in each set. Discuss with your partner and the class.

A.

quotient	remainder
factor	division

B.

multiple	multiplication
difference	factor

Wednesday: Rounding

Discuss the questions below. Prove it with numbers, words and/or pictures!

1. What are 3 numbers that round to 500?
2. What are 3 numbers that round to 7,000?
3. What are 3 numbers that round to 17,100?

Thursday: Number Talk

What are some ways to think about and show:

$$35 \times 7$$

Friday: What's the Question? (3 Read Protocol)

Read the problem 3 times with your class. The first time talk about the story. The second time discuss the numbers. The third time think of 2 questions that you could ask about this story. Write them down. Discuss with your classmates.

The jewelry store had 12 boxes with 5 gold rings in each box and 4 boxes with 20 silver rings in each box.

1.

2.

Week 14 Teacher Notes

Monday: Convince Me!

Students have to discuss the problem with a neighbor and figure out a way to discuss and defend their answer with numbers, words and pictures.

Tuesday: Frayer Model

This is a traditional model that gets students talking about a word. They should do it with a partner and then be ready to share out with the entire class.

Wednesday: Guess My Number

In this place value routine, students have to try and guess the target number by talking about the number.

Thursday: Number Talk

In this number talk students are looking at what it means to multiply by 2s. Discuss how we are doubling the number. Also look at the commutative property.

Friday: Regular Word Problem

Students use the picture prompt and write a story about the chocolate. It can be a repeated addition story or a multiplication story.

Week 14 Activities

Monday: Convince Me!

Discuss this problem with your neighbor. Convince them that it is true using numbers, words and models.

$$4 \times 9 = (2 \times 9) + (2 \times 9)$$

Tuesday: Frayer Model

Fill in the boxes based on the word.

Decompose

Definition	Examples
Give a Picture Example	**Non-examples**

Wednesday: Guess My Number

Read the clues. Guess the number.

A.	B.
I am a 2-digit number. I am greater than 62 and less than 77. I am a multiple of 3. I am not odd. The sum of my digits is less than 10. Who am I? 63, 66, 69, 72, 75	I am a 3-digit number. I am between 250 and 300. I am a multiple of 9. I am greater than 9×30. I am an even number. Who am I? 252, 261, 270, 279, 288, 297

Thursday: Number Talk

What are some ways to think about and show:

$$415 \div 7$$

A. First, talk about your estimate with your partner.

B. Then share your strategy for solving it.

Ideas: rectangular arrays, area models, repeated subtraction, partial quotients, properties of operations, and/or the relationship between multiplication and division.

Friday: Regular Word Problem

Read and solve the problem. Discuss your answer with a partner and then the whole group.

The chocolate shop is selling chocolates for $5 per pound. On Monday, they sold 38 pounds. On Tuesday, they sold 10 pounds more than they sold on Monday. On Wednesday, they sold 3 pounds less than they sold on Tuesday. How much did they sell altogether? How much money did they make?

Story:

Model:

Equation:

Week 15 Teacher Notes

Monday: Number Line It!

Students have to put the numbers in order on the number line.

Tuesday: Vocabulary Tic Tac Toe

Students play tic tac toe by writing down the definition or drawing a sketch.

Wednesday: Venn Diagram

Students reason about the items in the circles.

Thursday: British Number Talk

Give students about 5 minutes to work on their own or with partners to come up with some problems, hopefully from each category. Make sure that they stretch themselves. You don't want them to only choose the easy problem. Then, students should share what they did with the class.

Friday: Time Problem

Make up a word problem that tells what time he left and what time he came back.

Week 15 Activities

Monday: Number Line It!

Put these numbers in order on the number line. Be as exact as possible. Add 2 more fractions.

$\frac{2}{4}$ 1 $\frac{5}{4}$ $1\frac{1}{2}$

0 ⟷ 2

Tuesday: Vocabulary Tic Tac Toe

Play rock, paper, scissors to see who goes first. Then take turns, picking a square, saying what it means, drawing or writing something about the word on the side and then marking the word with an × or an o. Whoever gets 3 in a row first wins.

Say what is in each space.

$\frac{3}{4}$	numerator	
	$\frac{1}{2}$	denominator
equivalent	$\frac{2}{3}$	

Say what the fraction abbreviation stands for. Give an example of something that is measured in those units.

cm	yd	in.
l	ft	meter
ml	gm	kg

Wednesday: Venn Diagram

Fill in the Venn diagram.

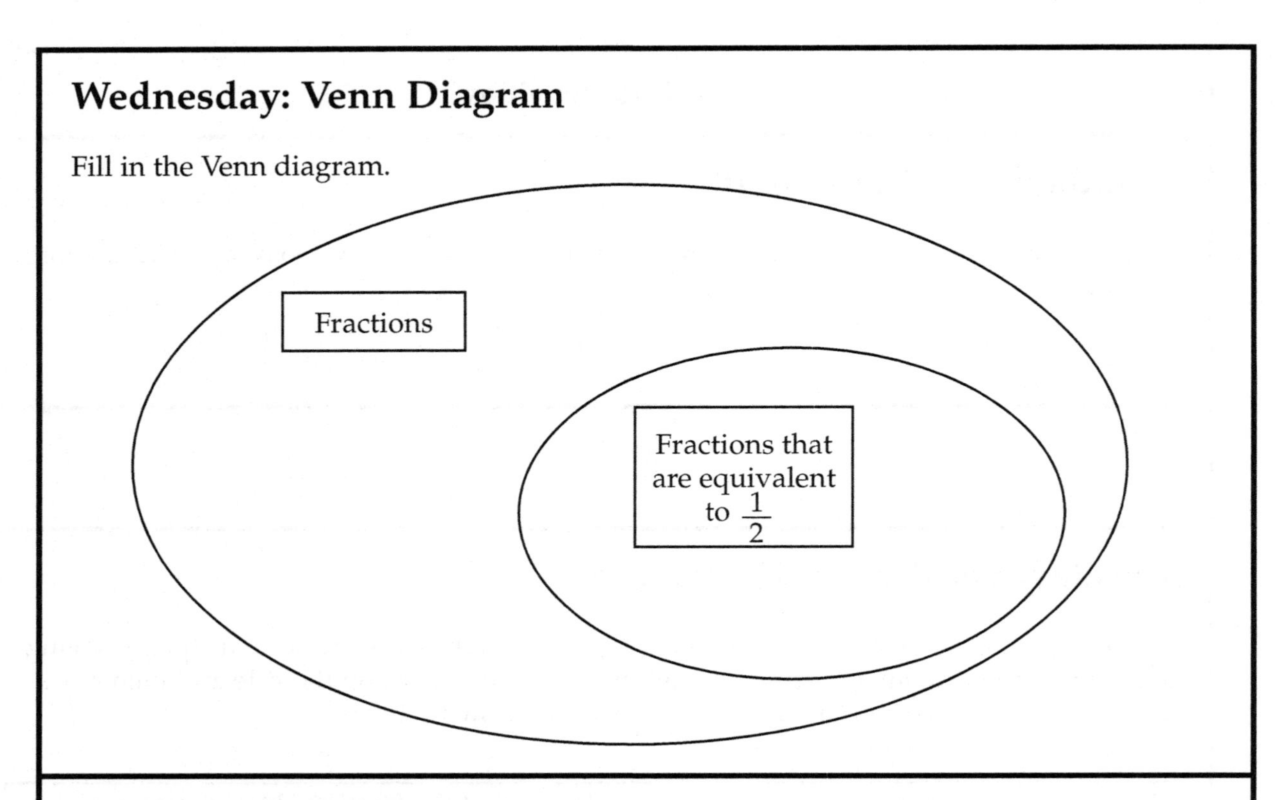

Thursday: British Number Talk

Pick a number from each circle. Make a multiplication problem. Write the problem under the way you solved it. For example, 10×10 is 100. I can do that in my head because I know it is a square number.

I can do it in my head.	I can do it with a model.	I can do it using a written strategy or algorithm.

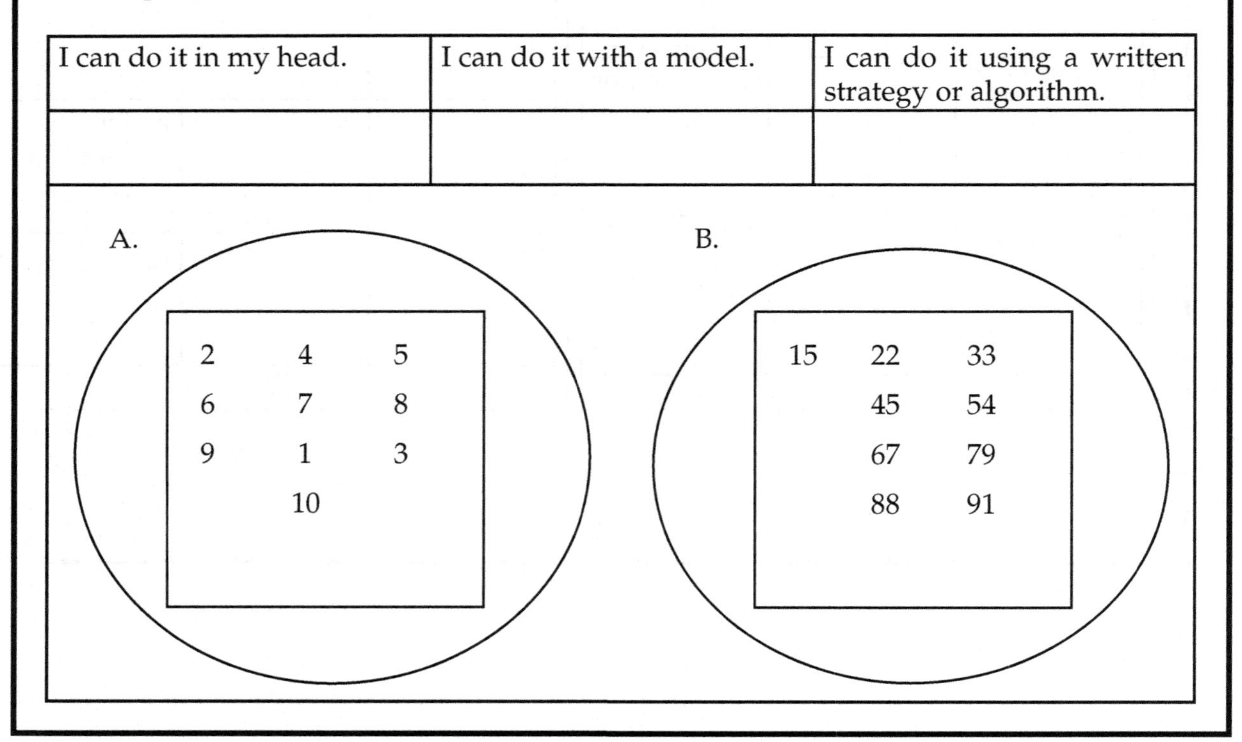

Friday: Time Problem

Lucas was gone for 1 hour and a half. At what time could he have left his house? At what time could he have returned?

1.

2.

Week 16 Teacher Notes

Monday: What Doesn't Belong?

When doing this routine have the students do the calculations (in their journals, on scratch paper or on the activity page). Then, have them share their thinking with a friend. Then, pull them back to the group.

Tuesday: 1-Minute Essay

Give students the designated part of the time to write and then share and then write again and then share out with the class.

Wednesday: Fraction of the Day

It is important that students are engaging in ongoing review of fractions. In this routine they are working with fourths/quarters, which they learned about in second grade.

Thursday: Number Talk

Give students about 5 minutes to work on their own or with partners to come up with some problems, hopefully from each category. Make sure that they stretch themselves. You don't want them to only choose the easy problem. Then, students should share what they did with class.

Friday: Sort That!

In this routine, students have to read and analyze fractions. They have to reason and decide how the fractions can be sorted. One way to approach this is to allow the students the opportunity to represent the fractions using models such as bar models or number lines and then discuss each of the fractions with their math partner. Do this for each of the fractions and then have the students discuss what the fractions have in common, or how they are different as they sort the fractions into groups. They defend their thinking to their partner and then the whole class comes back together and discusses their thinking.

Week 16 Activities

Monday: What Doesn't Belong?

A.

3×15	9×5
$88 - 43$	$100 - 45$

B.

6×2	$\frac{144}{12}$
$40 - 18$	$\frac{108}{9}$

Tuesday: 1-Minute Essay

(For 30 seconds) Write everything you can about multiplication. Use numbers, words and pictures.

(15 seconds) Now switch with a neighbor and add 1 thing to their list.

(15 seconds) Now add 1 more thing to your list.

Wednesday: Fraction of the Day

$$\frac{7}{12}$$

Word form	Rectangle model
Equivalent fraction	Plot on the number line. Write a fraction that is more than $\frac{7}{12}$ and one that is less than it.

Thursday: Number Talk

You multiply two numbers and the product is close to 1,000. What could have been the numbers you multiplied?

Friday: Sort That!

$\frac{3}{4}, \frac{4}{5}, \frac{1}{8}, \frac{4}{10}, \frac{1}{4}, \frac{1}{3}, \frac{6}{10}$

Sort the fractions into groups. What rule did you use to sort them?Explain your reasoning for sorting the fractions the way you did.

Week 17 Teacher Notes

Monday: Input/Output Table

Patterns are an important part of math in general and they are highlighted in many fourth-grade curriculums. Discuss these as a whole class and then give the students an opportunity to work with their math partner to create one of their own to share out.

Tuesday: Vocabulary Tic Tac Toe

These are quick partner energizers. Read all the words together. Then go! Students have 7 minutes to play the game. They do rock, paper, scissors to start. They take turns choosing a word and explaining it to their partner. Then, they have to do a sketch or something to show they understand the word. Everybody should play the first game, if they have time, they can play the next one.

It is important to call everyone back together at the end and talk about the vocabulary. Briefly go over the vocabulary, this is all second-grade vocabulary.

Wednesday: What Doesn't Belong?

When doing this routine have the students do the calculations (in their journals, on scratch paper or on the activity page). Then, have them share their thinking with a friend. Then, pull them back to the group. The focus is to get students to talk about numbers in general and rounding.

Thursday: Number Strings

Students should discuss the string as a whole class. You want the emphasis to be on the relationships between the numbers. In this string we are working on the relationship between 2s and 4s. With practice, students do this naturally.

Friday: Fill in the Problem!

In this routine students work on their own and make their own problem. When they are done they share it with their math partner to see if it makes sense and they did it correctly. After that, they should be ready to share their thinking with the whole class.

Week 17 Activities

Monday: Input/Output Table

Look at the tables. Figure out the rule. Fill in the empty spaces.

What's the rule? A.		What's the rule? B.		Make your own…. C.	
In	Out	In	Out	In	Out
4	28	20	4		
5	?	40	8		
?	42	15	?		
7	49	5	?		
8	?	?	0		
0	?	?	2		

Tuesday: Vocabulary Tic Tac Toe

Play rock, paper, scissors to see who goes first. Then take turns, picking a square, saying what it means, drawing or writing something about the word on the side and then marking the word with an × or an o. Whoever gets 3 in a row first wins.

partial quotient	line plot	picture graph
partial sum	multiple	partial differences
factor	bar graph	partial product

rounding	estimate	multiplication
equal group	angle	vertex
array	area	perimeter

Wednesday: What Doesn't Belong?

Look at the boxes. Pick the one that doesn't belong.

A.

$\frac{2}{3}$	$\frac{4}{6}$
$\frac{6}{8}$	$\frac{8}{12}$

B.

$\frac{1}{4}$	$\frac{2}{8}$
$\frac{3}{12}$	$\frac{4}{4}$

Thursday: Number Strings

Discuss what you notice. How are these problems related?

10×2 5×2 10×6 5×6	10×9 5×9 10×8 5×8	10×9 20×9 10×11 20×11

Friday: Fill in the Problem!

Pick the numbers from those given and then solve the problem.

Mike had ______ marbles. He put _______ in an equal amount in boxes. How many boxes did he use?

(12 or 15) (2, 3 or 4)

How many boxes did he have left over?

Week 18 Teacher Notes

Monday: Missing Number

Students have to fill in the missing number.

Tuesday: What Doesn't Belong?

In this routine, students are thinking about the relationship between the words. First have the students read all the words to their partner and discuss the definitions. Then have them decide what they think doesn't belong and why. After everyone has thought about it, discuss it as a class.

Wednesday: Bingo

Students have to reason about fractions.

Thursday: Number Talk

This is a typical number talk where students are thinking about the ways in which they can solve the problem.

Friday: Model It

The students have to model the word problem and then solve it.

Week 18 Activities

Monday: Missing Number

Fill in the missing numbers.

A. $35 \div ? = 7$

B. $5 \times \_\_ \times \_\_ = 100$

C. $35 + \_\_\_ + \_\_\_ = 90$

D. $55 = 100 - ?$

Tuesday: What Doesn't Belong?

Look at the boxes. Pick the one that doesn't belong.

A.

km	m
ml	cm

B.

inches	yards
ounces	feet

Wednesday: Bingo

Put these numbers on your bingo board. Do not put them in the order that they are written. Scatter them in different places. Your teacher will call them out. Mark them on your board. Whoever gets 4 in a row horizontally, vertically or as a postage stamp (4 in a corner) first wins.

$\frac{1}{2}$ $\frac{3}{4}$ $\frac{5}{7}$ $\frac{3}{9}$ $\frac{4}{4}$ $\frac{16}{8}$ $\frac{9}{8}$ $\frac{30}{5}$
$\frac{2}{4}$ $\frac{4}{6}$ $\frac{2}{10}$

$\frac{3}{4}$ $\frac{1}{7}$ $\frac{2}{6}$ $\frac{1}{8}$ $\frac{2}{3}$ $\frac{4}{5}$ $\frac{1}{2}$
$\frac{7}{9}$ $\frac{8}{10}$ $\frac{11}{12}$ $\frac{5}{10}$ $\frac{4}{2}$

Thursday: Number Talk

What are some ways to think about $\frac{4,608}{9}$

Ideas: rectangular arrays, area models, repeated subtraction, partial quotients, properties of operations, and/or the relationship between multiplication and division.

Friday: Model It

Read the problem. Solve in 2 different ways.

Marvin has 10 marbles. He has 2 times as many as his brother. How many does his brother have? How many do they have altogether?

Model this problem in 2 different ways:

Way 1	Way 2

Week 19 Teacher Notes

Monday: Break It Up!

In this routine, students work on modeling the distributive property. This is an important skill that students need to know and the way they learn to do it is across time.

Tuesday: 1-Minute Essay

Give students some designated time to write and then share and then write again and then share out with the class.

Wednesday: How Many More to

This is a place value routine and students are thinking about number relationships.

Thursday: British Number Talk

Give students about 5 minutes to work on their own or with partners to come up with some problems, hopefully from each category. Make sure that they stretch themselves. You don't want them to only choose the easy problem. Then, students should share what they did with class.

Friday: What's the Story? (Here's the Model)

In this routine, students are given the model (in this case the tape diagram) and they have to tell a story that matches the model. There are many ways to do this. Students can talk with their math partner first and then share their thinking with the class. Or, the whole class can brainstorm what the possibilities are and then the students work on their own story after hearing several possibilities.

Week 19 Activities

Monday: Break It Up!

3×7

Sketch it!	Break it apart!
	$3 \times 7 =$ (____ $\times$ _____) + (_____ $\times$ ______)

Tuesday: 1-Minute Essay

(For 30 seconds) Write everything you can about division. Use numbers words and pictures.

(15 seconds) Now switch with a neighbor and add 1 thing to their list.

(15 seconds) Now add 1 more thing to your list.

Wednesday: How Many More to

Start at 50 … Get to 450 … skip Count by 50s.
Start at 79 … Get close to 100 … Skip count by 2s.
Start at 1050 … Get close to 3000 … Skip count by 150s.
Start at 275 … Get to 500 … Skip count by 25s.

Thursday: British Number Talk

Pick a number from each circle. Make an addition problem. Write the problem under the way you solved it. For example, 3,182 + 138. I broke it apart with the numbers. I did 3,000 + 200 + 110 + 10. I did it with the partial sums strategy.

I can do it in my head.	I can do it with a model.	I can do it using a written strategy or algorithm.

A.

3,182
1,429 1,657
3,785 2,846
1,993 5,234
4,578 3,501
77

B.

138 219 59
263 714
65 170 101
22 100 99
10 178 87 43

Friday: What's the Story? (Here's the Model)

Tell a story that matches the model.

3	3	3	3		
3	3	3	3	3	3

Week 20 Teacher Notes

Monday: True or False?

In this routine, students think about and decide whether or not the equations are true or false. They should do it with a partner and then discuss it as a whole class.

Tuesday: Vocabulary Brainstorm

Students have to think and then write about the word in all of the bubbles. They first share their thinking with a partner and then share their thinking with the class.

Wednesday: Find and Fix the Error

Students have to discuss how this problem is incorrect. They need to talk about what is wrong, as well as how to fix it.

Thursday: Number Talk

Students have to talk about decomposing $\frac{12}{10}$. They can use various models to scaffold their thinking.

Friday: What's the Story? (Here's the Model)

In this routine, students are given the model (in this case the tape diagram) and they have to tell a story that matches the model. There are many ways to do this. Students can talk with their math partner first and then share their thinking with the class. Or, the whole class can brainstorm what the possibilities are and then the students work on their own story after hearing several possibilities.

Week 20 Activities

Monday: True or False?

	True or False?
$\frac{5}{5} + \frac{3}{5} = 1\frac{3}{5}$	
$\frac{16}{8} = \frac{1}{2}$	
$\frac{3}{4} + \frac{3}{4}$ is greater than $\frac{6}{8}$	
$1\frac{1}{2} + 1\frac{1}{2}$ is less than 3	
Make your own!	

Tuesday: Vocabulary Brainstorm

In each thought cloud write or draw something that has to do with equivalent.

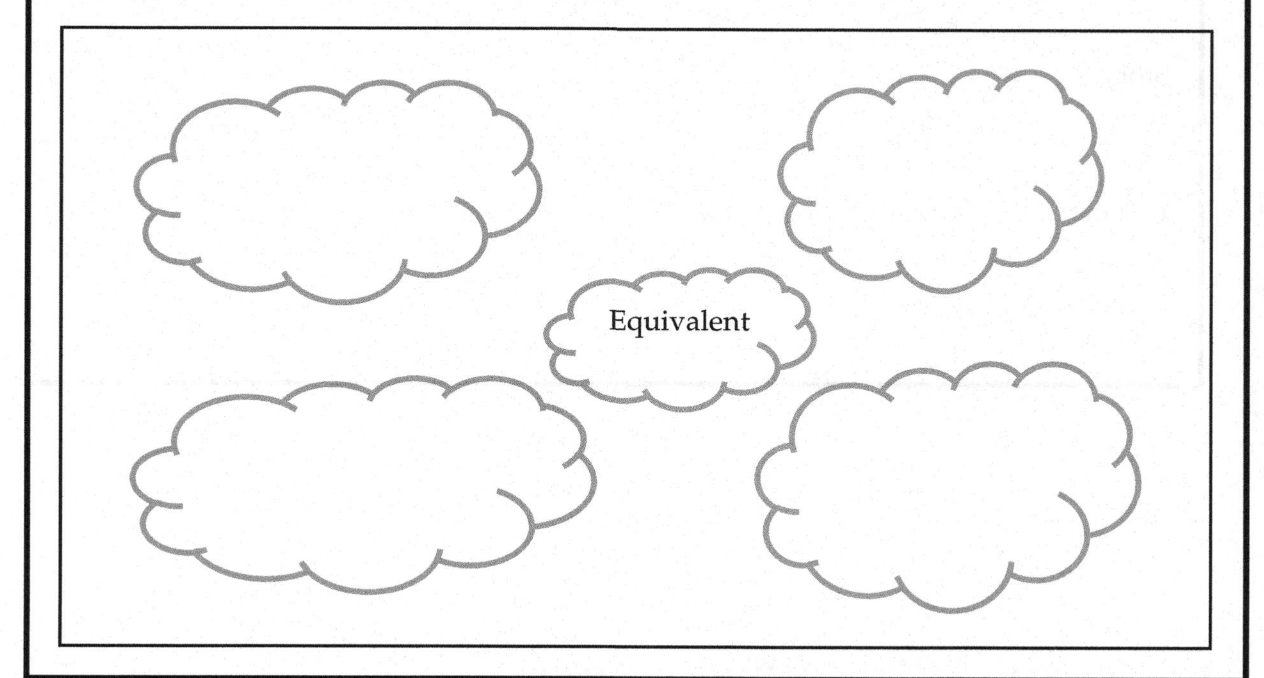

Wednesday: Find and Fix the Error

Luke did this…

$4 \times \frac{1}{2} = \frac{4}{8}$

What did he get wrong? How can he fix it?

Thursday: Number Talk

What are some ways to think about decomposing this fraction:

$$\frac{12}{10}$$

Friday: What's the Story? (Here's the Model)

Tell a story to match this tape diagram.

5	5	5	5
5			

Story:

Equation:

Week 21 Teacher Notes

Monday: True or False?

Students have to think about whether or not the problems are true or false. They have to defend their thinking to their math partner and then be ready to discuss it in the whole group.

Tuesday: Vocabulary Bingo

Students have to put the words in different spaces. Teacher calls the words by stating a definition or showing a picture. When the teacher gives the definition or shows a picture, students cover the correct word. Whoever gets 4 in a row vertically, horizontally or 4 corners first wins.

Wednesday: Number Bond It!

Students have to show how to break apart a gallon in 3 different ways.

Thursday: British Number Talk

Students pick a number from each circle. Multiply the numbers. Decide how they will solve it and write that expression under the correct category.

Friday: What's the Story?

Students are given the answer. They have to write a division story with a remainder of 4.

Week 21 Activities

Monday: True or False?

Read the problems. Decide if they are true or false. Discuss why or why not.

A. $2 \times 2 \times 7 = 4 \times 7$	B. $60 \div 5 = (30 \div 5) + (30 \div 5)$	C. Make your own and share it out.

Tuesday: Vocabulary Bingo

Put these words on your bingo board. Do not put them in the order that they are written. Scatter them in different places. Your teacher will call them out. Mark them on your board. Whoever gets 4 in a row horizontally, vertically or as a postage stamp (4 in a corner) first wins.

Words: area, tape/strip diagram, classify, horizontal, centimeter, partial quotient, partial product, gram, kilogram, quadrilateral, parallelogram, vertical.

Write a word in each space.

Wednesday: Number Bond It!

Show how to break apart a gallon in 3 different ways!

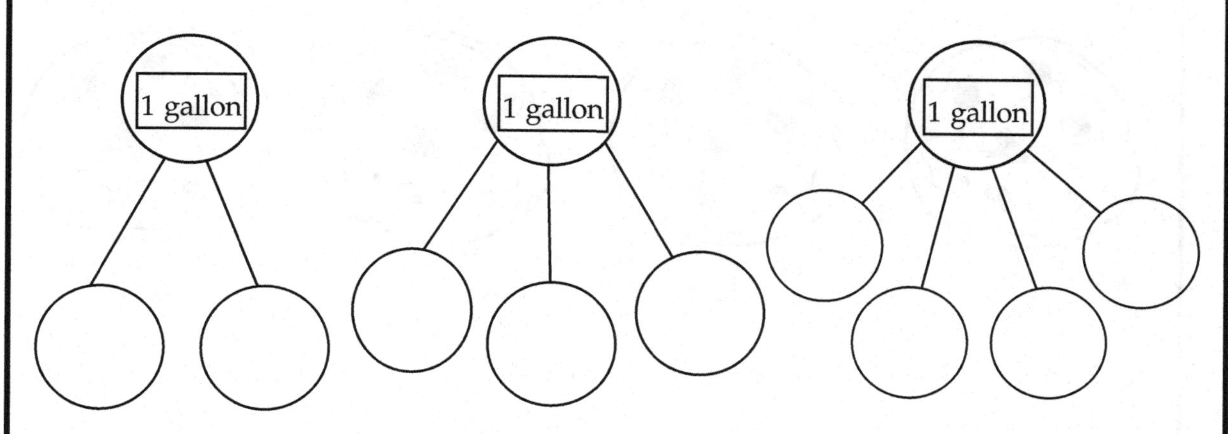

Thursday: British Number Talk

Pick a number from each circle. Make a multiplication problem. Write the problem under the way you solved it.

I can do it in my head.	I can do it with a model.	I can do it using a written strategy or algorithm.

A.

10 12
15 20 11

B.

11 12 13
14 15 16
17 18
19 10

Friday: What's the Story?

The remainder is 4 cookies. It was a division story. What was the story?

Week 22 Teacher Notes

Monday: Missing Numbers

Students have to talk about what happens when you subtract fractions with like denominators.

Tuesday: Frayer Model

Students must fill in the squares based on the word.

Wednesday: Number of the Day

Fill in the boxes to represent the number in different ways.

Thursday: Number Talk

In this number talk you want the students to discuss their thinking with strategies and models. Ask students about the strategies that they might use.

Friday: What's the Question? Here's the Graph

Students have to read the graph and then make up a story about what the data represents and tell it to their partner. They have to explain their thinking and be ready to share their thinking to the whole group.

Week 22 Activities

Monday: Missing Numbers

Talk about possible numbers that could be the numerators.

A.

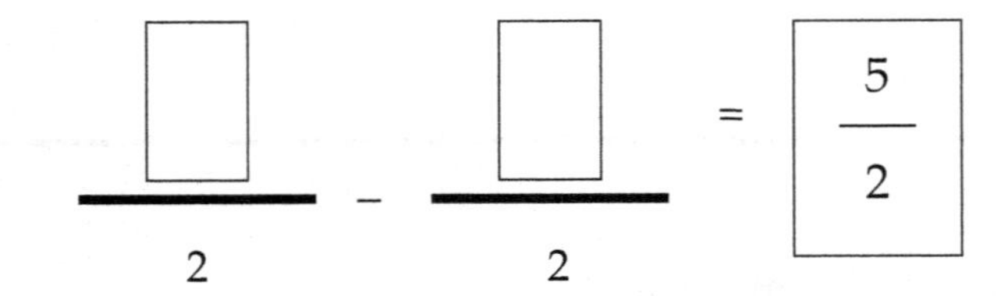

$$\frac{\square}{2} - \frac{\square}{2} = \frac{5}{2}$$

What happens when you subtract fractions with like denominators?

Tuesday: Frayer Model

Fill in the boxes based on the word.

Remainder

Definition	Examples
Give a Picture Example	**Non-examples**

Wednesday: Number of the Day

Fill in the boxes based on the number.

82,091

Word form	Expanded form	100 more is ______. 100 less is _______.
____ is greater than _____ ____ is less than _______	____ + ____ = 82,091	____ – ____ = 82,091

Plot it on the number line. Also put 2 numbers that are larger than the number and 2 numbers that are smaller than the number.

Thursday: Number Talk

What are some ways to think about and represent:

346×4

Friday: What's the Question? Here's the Graph

Here's the graph.	What's the story behind the graph? Make up a story and tell it to your partner. Explain your thinking. Be ready to share your thinking to the whole group.

$\frac{1}{2}$	1	$1\frac{1}{2}$	2	$2\frac{1}{2}$
x x	x x x x	x x x x x x	x x x	x x x x x x x

Week 23 Teacher Notes

Monday: Magic Square

Students have to fill in the squares with numbers so that all directions end in a sum of 34.

Tuesday: Frayer Model

Students have to fill in the different boxes to represent the word.

Wednesday: Missing Number

Students have to discover the pattern and then fill in the missing numbers.

Thursday: Number Talk

Students discuss ways to solve the expression.

Friday: Model It

Students have to model the problem with a tape diagram and solve it.

Week 23 Activities

Monday: Magic Square

Sum is 34

Fill in all the spaces so that each row and column and diagonal totals to 34.

1		14	4
12	6		9
	3		16

Tuesday: Frayer Model

Fill in the boxes based on the word.

Area

Definition	Examples
Give a Picture Example	**Non-examples**

Wednesday: Missing Number

Discover the pattern. Fill in the missing numbers.

A. 72, ____, _____, 108, 120, _____, 144

B. 380, 160, _______, _______, _______, 10, _____

C. Make your own pattern:

_____, _______, _______, ________, ________, _________, _________

Thursday: Number Talk

What are some ways to solve 62 ÷ 9?

Friday: Model It

Read the problem. Model it and write the equation.

In the aquarium there were 15 fish. There were 2 times as many turtles as fish. How many turtles were there? How many animals were there altogether?

Model it with a tape/strip diagram.

Equation:

Week 24 Teacher Notes

Monday: Input/Output Table

Students have to create an input output table where the rule is to multiply by 11.

Tuesday: 1-Minute Essay

Students have 1 minute to write everything they can about mixed numbers using numbers, words and pictures.

Wednesday: Bingo

Students put the numbers on the bingo board in a different order than they are listed. The teacher calls them out using different ways to represent them and describe them. Whoever gets 4 in a row horizontally, vertically or as a postage stamp (4 in a corner) first wins.

Call them in a different order.

- Cover a prime number.
- Cover a composite number.
- Cover a factor of 25.
- Cover a factor of 16.
- Cover the number that when you multiply it, it is always the other number.
- Cover a factor of 18.
- Cover a multiple of 2.
- Cover a factor of 11.
- Cover a factor of 7.
- Cover a factor of 3.
- Cover a factor of 5.
- Cover a multiple of 3.
- Cover a multiple of 4.
- Cover a multiple of 5.
- Cover a multiple of 6.
- Cover the quotient of $\frac{18}{9}$.
- Cover the quotient of $\frac{18}{3}$.
- Cover the quotient of $\frac{18}{6}$.
- Cover the quotient of $\frac{12}{3}$; $\frac{12}{4}$; $\frac{12}{6}$.
- Cover the quotient of $\frac{20}{4}$; $\frac{20}{5}$; $\frac{20}{2}$; $\frac{20}{10}$.

Thursday: British Number Talk

Students pick a number from each circle. Divide them. Decide how they will solve it and write that expression under the correct category.

Friday: Model It

Students have to read, model and solve the word problem.

Week 24 Activities

Monday: Input/Output Table

Create an input output table where the rule is to multiply by 11.

In	Out

Tuesday: 1-Minute Essay

You have 1 minute to write everything you know about mixed numbers.

Go!

Wednesday: Bingo

Put these numbers on your bingo board. Do not put them in the order that they are written. Scatter them in different places. Your teacher will call out clues. Mark them on your board. Whoever gets 4 in a row horizontally, vertically or as a postage stamp (4 in a corner) first wins.

1, 2, 3, 4, 5, 6, 7, 8, 9, 10, 11, 12, 13, 14, 15, 16

Thursday: British Number Talk

Pick a number from each circle. Divide them. Decide how you will solve it and write that expression under the title.

I can do it in my head.	I can do it with a model.	I can do it using a written strategy or algorithm.

A.

28 100 182
240 125 74
400 65 486
147 23 451

B.

1 2 3
4 5 6
7 8 9

Friday: Model It

Read the problem. Model it.

At the butterfly museum there were 40 butterflies in the garden. This was 5 times as many butterflies as ladybugs. How many ladybugs were there? How many insects where there altogether?

Week 25 Teacher Notes

Monday: What Doesn't Belong?

When doing what doesn't belong, have the students do the calculations (in their journals, on scratch paper or on the activity page). Then, have them share their thinking with a friend. Then, pull them back to the group.

Tuesday: Vocabulary Tic Tac Toe

These are quick partner energizers. Read all the words together. Then go! Students have 7 minutes to play the game. They do rock, paper, scissors to start. They take turns choosing a word and explaining it to their partner. Then, they have to do a sketch or something to show they understand the word. Everybody should play the first game, if they have time, they can play the next one.

It is important to call everyone back together at the end and talk about the vocabulary. Briefly go over the vocabulary.

Wednesday: Fraction of the Day

Students have to fill in the boxes to represent the fraction.

Thursday: Number Talk

In this number talk you want the students to discuss their thinking with strategies and models. Ask students about the strategies that they might use.

Friday: Model It

Students have to read, model and solve the word problem.

Week 25 Activities

Monday: What Doesn't Belong?

Look at each set. Decide which expression does not belong in each set. Discuss with your partner and the class.

A.

$144 \div 12$	$6 \times ? = 72$
$132 \div 11$	6×3

B.

$4 \times \frac{1}{2}$	$\frac{1}{2} + 1\frac{1}{2}$
$2 \times \frac{1}{2}$	$2\frac{1}{2} - \frac{1}{2}$

Tuesday: Vocabulary Tic Tac Toe

Play rock, paper, scissors to see who goes first. Then take turns, picking a square, saying what it means, drawing or writing something about the word on the side and then marking the word with an × or an o. Whoever gets 3 in a row first wins.

operation	mixed number	multiple
expression	factor	right angle
equation	equivalent	angle

length	weight	metric
pounds	ounces	customary
capacity	height	mass

Wednesday: Fraction of the Day

Fill in the boxes based on the fraction.

$$1\frac{1}{2}$$

Word form	Picture form
Plot it on a number line	Is it greater than or less than $\frac{7}{6}$? How do you know?

Thursday: Number Talk

You add two fractions and the sum is $\frac{4}{6}$. What could the fractions be? How do you know?

What if you add three fractions, but the sum is still $\frac{4}{6}$. What could the fractions be?

Friday: Model It

Read the problem and model it. Solve the problem.

Mr. Lucas had 1 piece of wood that was 25 ft long. He had another piece of wood that was 5ft long. How much longer was the first piece of wood than the second piece of wood?

Model with a tape/strip diagram:

Week 26 Teacher Notes

Monday: Open Array Puzzle

Students have to figure out the partial quotients and fill them in the open array.

Tuesday: Vocabulary Match

Students have to match the words and the definition.

Wednesday: Guess My Number

Students have to read the clues and guess the mystery fraction.

Thursday: Open Array Puzzle

Students have to look at the puzzle and fill in the missing numbers.

Friday: What's the Story?

Students have to write a story to match the model.

Week 26 Activities

Monday: Open Array Puzzle

$\frac{457}{8}$

________ + ________

?	?

Answer is _______ remainder 1

Tuesday: Vocabulary Match

Match the words and the definitions.

Acute angle	An angle measuring 90 degrees
Obtuse angle	An angle less than 90 degrees
Straight line	An angle measuring 180 degrees
Right angle	An angle greater than 90 degrees and less than 180

Wednesday: Guess My Number

Use the clues to figure out the mystery fraction.

I am a fraction that is greater than $\frac{1}{2}$. I am less than $\frac{3}{4}$. My numerator is odd. My denominator is even. Who am I?	$\frac{3}{12}$ $\frac{4}{8}$ $\frac{5}{8}$ $\frac{9}{6}$

Thursday: Open Array Puzzle

Fill in the boxes and write the partial products.

$12 \times 17 =$

	10 + 7	
10		
2		

Friday: What's the Story?

Look at the model. Make up a story that matches the model. Discuss with your partner and then the whole group.

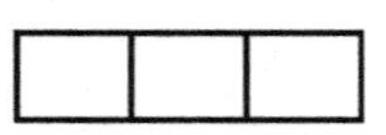

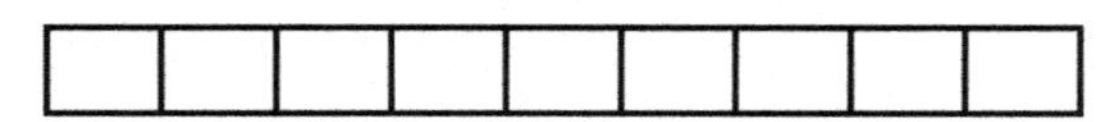

Week 27 Teacher Notes

Monday: 3 Truths and a Fib

Read the boxes. 3 are true and one is false. Which one is false?

Tuesday: Convince Me!

Students have to convince each other that these are not prime numbers. They can use some of the suggested phrases.

This is true! I can prove it with . . .
This is the same because . . .
I am going to use _________ to show my thinking.
I am going to defend my answer by _________.

Wednesday: Pattern/Skip Counting

Describe and complete that pattern.

Thursday: Number Talk

You want students to think about different strategies to solve it as well as different models to represent it.

Friday: Equation Match

Students have to figure out the missing number. They also have to match the problem with the equation.

Week 27 Activities

Monday: 3 Truths and a Fib

Figure out which one is not true.

7 is not composite	8 is not prime
20 is composite	12 is prime

Tuesday: Convince Me!

Convince me that these are not prime numbers.

4 9 15

Prove it with numbers, words and/or pictures!

You have to defend your thinking with numbers, words and pictures. Discuss this with your math partner and then the whole group.

This is true! I can prove it with . . .
This is the same because . . .
I am going to use __________ to show my thinking.
I am going to defend my answer by __________.

Wednesday: Pattern/Skip Counting

1. $\frac{1}{4}$, $\frac{2}{4}$, $\frac{3}{4}$ ___, ___, _____
2. $\frac{2}{3}$, $1\frac{1}{3}$, 2, ____, ____, ____, $4\frac{2}{3}$
3. Make your own:
   ____, ____, ____, ____, ____, ____, ____, ____,

Thursday: Number Talk

What are some ways to solve $4 \times \frac{1}{2}$?

How can you do this with repeated addition?

How can you represent this on a number line?

Friday: Equation Match

Figure out the missing number. Match the equation to the problem. Describe how you know it is the correct answer.

$$\frac{4}{5} - \_\_ = \frac{1}{5}$$

Problem A	Problem B
Aunty Mary had $\frac{4}{5}$ cup of butter. She used $\frac{2}{5}$. How much does she have left?	Aunty Mary had $\frac{4}{5}$ cup of butter. She used some. Now, she has $\frac{1}{5}$ left. How much did she use?

Week 28 Teacher Notes

Monday: Why Is It Not?

Students look at each problem and think about why the given answer is not correct. They have to discuss it and then explain their thinking to the class.

Tuesday: 1-Minute Essay

Students have to write everything they know about angles. They should some of these words in their quickwrite.

Wednesday: Guess My Number

Students read the clues and figure out which one is the missing fraction.

Thursday: British Number Talk

Students pick a number from each circle. Make a multiplication problem. Write the problem under the way you solved it. For example, "I can do that in my head. I can multiply $\frac{1}{2} \times \frac{2}{3}$ and I know that is $\frac{2}{6}$."

Friday: What's the Question? (3 Read Protocol)

Read the problem 3 times chorally with your class.
The first time talk about the problem.
The second time talk about the numbers.
The third time, talk about what questions you could ask.

Week 28 Activities

Monday: Why Is It Not?

Think about it and share it with a partner. Share with the class.

$\frac{6}{8}$ + ____ = $\frac{8}{8}$

Why is it not $\frac{14}{8}$?

Tuesday: 1-Minute Essay

Write everything you know about angles.

Vocabulary bank: more than, less than, greater than, degrees, obtuse, straight, right angle.

Wednesday: Guess My Number

Read the clues and figure out which one is the missing fraction.

I am a fraction that is greater than $\frac{6}{5}$. I am less than $\frac{4}{2}$. My numerator is odd. My denominator is odd. Who am I?	$\frac{7}{5}$ $\frac{3}{4}$ $\frac{2}{3}$ $\frac{6}{6}$ $\frac{11}{5}$

Thursday: British Number Talk

Pick a number from each circle. Make a multiplication problem. Write the problem under the way you solved it.

I can do it in my head.	I can do it with a model.	I can do it using a written strategy or algorithm.

A.

1 2 3

4 5 6

7 8 9

10 0

B.

$\frac{1}{2}$ $\frac{2}{3}$ $\frac{3}{4}$

$\frac{4}{5}$ $\frac{5}{10}$

$\frac{6}{8}$ $\frac{2}{6}$ $\frac{4}{12}$

$\frac{0}{6}$

Friday: What's the Question? (3 Read Protocol)

Read the problem 3 times with your class. The first time talk about what the problem is about. The second time talk about what the numbers mean. The third time talk about at least 2 questions you could ask about this story. Write them down. Discuss with your classmates.

Grandma Betsy made 4 pies. She needed to put $\frac{2}{4}$ of a cup of butter in each one. She also put $\frac{3}{4}$ cup of sugar and $\frac{1}{4}$ cup of milk in each pie she made.

Week 29 Teacher Notes

Monday: Guess My Number

Students have to read the clues and guess the number.

Tuesday: Vocabulary Bingo

Students have to put these words on the bingo board in different spaces. They should not put them in the order they appear. They listen to the teacher call out the definitions or show a picture. When you hear it cross out the word. Whoever gets 4 in a row first wins.

Wednesday: Bingo

Students have to put these fractions on the bingo board in different spaces. They should not put them in the order they appear. They listen to the teacher call out the definitions or show a picture. When you hear it cross out the word. Whoever gets 4 in a row first wins.

$\frac{1}{2}, \frac{3}{4}, \frac{2}{3}, \frac{4}{5}, \frac{9}{7}; \frac{2}{4}; \frac{6}{8}; \frac{11}{12}; \frac{7}{7}; \frac{3}{3}; \frac{1}{3}; \frac{1}{8}; \frac{3}{5}; \frac{4}{6}; \frac{6}{2}; \frac{4}{2}; \frac{0}{1}; \frac{1}{1}$

Students have to pick 12 of these numbers to put on the board in any order.

Call out numbers by equivalent fractions, show the word form, show different models (area, set, linear, measurement) and say addition, subtraction and multiplication equations where these fractions are the answer.

Thursday: Number Talk

Students have to discuss this missing addition puzzle. They work together to figure it out.

Friday: Make Your Own Problem!

Students have to write a word problem where the answer is $\frac{5}{6}$ of a cup of butter.

Week 29 Activities

Monday: Guess My Number

Read the clues. Use them to figure out the mystery number.

Guess my number.
I am thinking of a number that is greater than 12×7.
I am thinking of a number that is less than $150 - 60$.
It is not an odd number.
The sum of the digits is greater than 14.
What is my number?

Tuesday: Vocabulary Bingo

Put these words on your bingo board. Do not put them in the order that they are written. Scatter them in different places. Your teacher will call them out. Mark them on your board. Whoever gets 4 in a row horizontally, vertically or as a postage stamp (4 in a corner) first wins.

Words: mixed number, product, angle, acute, obtuse, straight, right angle, parallel, perpendicular, ray, intersecting, degree, circle, line, end point, line segment.

Write a word in each space.

Wednesday: Bingo

Put these numbers on your bingo board. Do not put them in the order that they are written. Scatter them in different places. Your teacher will call them out. Mark them on your board. Whoever gets 4 in a row horizontally, vertically or as a postage stamp (4 in a corner) first wins.

$\frac{1}{2}, \frac{3}{4}, \frac{2}{3}, \frac{4}{5}, \frac{9}{7}; \frac{2}{4}; \frac{6}{8}; \frac{11}{12}; \frac{7}{7}; \frac{3}{3}; \frac{1}{3}; \frac{1}{8}; \frac{3}{5}; \frac{4}{6}; \frac{6}{2}; \frac{4}{2}; \frac{0}{1}; \frac{1}{1}$

Pick 16 of these numbers to put on the board in any order.

Thursday: Number Talk

Discuss the problem. What is missing? How do you know?

$$\begin{array}{r} \square\;9\;8 \\ +\;5\;\square\;3 \\ \hline 1{,}17\;\square \end{array}$$

Friday: Make Your Own Problem!

Make up a word problem where the answer is $\frac{5}{6}$ of a cup of butter.

Week 30 Teacher Notes

Monday: 2 Arguments

Students have to read both arguments and decide which one they agree with. They need to justify their thinking with numbers, words and/or pictures.

Tuesday: Vocabulary Bingo

Students have to put these words on the bingo board in different spaces. They should not put them in the order they appear. They listen to the teacher call out the definitions or show a picture. When they hear it, they cross out the word. Whoever gets 4 in a row first wins.

The teacher calls out the vocabulary by showing pictures and giving definitions.

Wednesday: Money Combinations

Students have to show and discuss 3 different ways to make the designated amount.

Thursday: British Number Talk

Students pick a number from each circle. They make a division problem. Then they write the problem under the way they solved it. For example, "I can do that in my head. 114 divided by 2 in my head. I know that 100 divided by 2 is 50 and 14 divided by 2 is 7 so it is 57.

Friday: Picture That!

Students have to tell a fraction story about the box of donuts.

Week 30 Activities

Monday: 2 Arguments

Read through the problem. Figure out who is correct. Discuss your thinking.

$42 \div \_\_ = 6$

John said the answer was 48.

Maria said the answer was 7.

Who do you agree with?

Why do you disagree with John/Maria?

Tuesday: Vocabulary Bingo

Put these words on your bingo board. Do not put them in the order that they are written. Scatter them in different places. Your teacher will call them out. Mark them on your board. Whoever gets 4 in a row horizontally, vertically or as a postage stamp (4 in a corner) first wins.

Words: pint, ounce, metric, customary, cup, quart, gallon, liter, gram, kilogram, kilometer, kiloliter milliliter, centimeter, millimeter, tablespoon.

Write a word in each space.

Wednesday: Money Combinations

Raul had $4.12 Show 3 different ways he could have had this money.

Way 1	Way 2	Way 3

Thursday: British Number Talk

Pick a number from each circle. Make a division problem. Then, decide how you are going to solve it. Write the problem under the way you solved it.

I can do it in my head.	I can do it with a model.	I can do it using a written strategy or algorithm.

512 114
416 218 321
615 724 840
100

1 3 5
2 4 6
8
9 7

Friday: Picture That!

Tell a fraction story about this box of donuts.

Week 31 Teacher Notes

Monday: 3 Truths and a Fib

Students have to read the statements and decide which 3 are true and which one is not. They then discuss with their neighbor which one is false. They must explain their thinking to their neighbor and then to the class.

Tuesday: Vocabulary Brainstorm

Students fill in the clouds with numbers, words and pictures to discuss the target word.

Wednesday: How Many More to

Students have to say how many more from the given number to the designated number.

Thursday: Number Talk

This is a typical number talk where students are thinking about the ways in which they can solve this subtraction problem. You want students to think about partial sums, counting up, and compensation. Students are focusing on different ways.

Friday: Fill in the Problem!

Students fill in the word problem and then solve it.

Week 31 Activities

Monday: 3 Truths and a Fib

Which one is false? Why? Explain to your neighbor and then the group.

210 = 21 tens	210 = 10 tens and 110 ones
210 = 10 tens and 21 ones	210 = 20 tens and 10 ones

Tuesday: Vocabulary Brainstorm

In each thought cloud write or draw something that has to do with the vocabulary word.

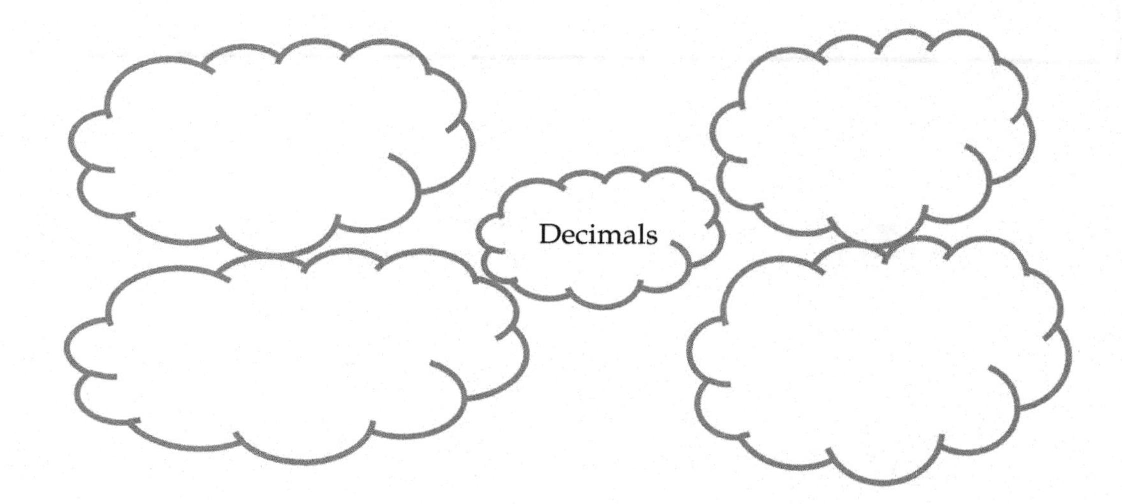

Wednesday: How Many More to

Start at $\frac{1}{4}$ … Get to $\frac{1}{2}$ …

Start at $\frac{2}{3}$ … Get to $\frac{9}{6}$ …

Start at $\frac{2}{10}$ … Get close to $\frac{4}{5}$ …

Start at $\frac{1}{3}$ … Get to $\frac{9}{12}$ …

Thursday: Number Talk

What are some ways to solve $\frac{3}{10} + \frac{4}{100}$?

Friday: Fill in the Problem!

Fill in and solve.

Jan ran ________ miles for ________ days in a row.

$\frac{1}{2}, \frac{1}{4}, \frac{1}{3}$ 2, 3, or 4

How far did she run in total?

Week 32 Teacher Notes

Monday: Reasoning Matrix

Students read the clues and then figure out which shapes belong to which students.

Tuesday: Vocabulary Bingo

Students have to put these words on the bingo board in different spaces. They should not put them in the order they appear. They listen to the teacher call out the definitions or show a picture. When they hear it, they cross out the word. Whoever gets 4 in a row first wins.

The teacher calls out the vocabulary by showing pictures and giving definitions.

Wednesday: Greater than, Less Than, in Between

In this routine the students put fractions into the boxes based on the descriptions.

Thursday: British Number Talk

Students pick a number from each circle. Students add the fractions. They decide how they will solve it and write that expression under the title.

Friday: What's the Story? Here's the Model

Students have to look at the tape diagram below and write a problem that matches the model.

Week 32 Activities

Monday: Reasoning Matrix

	Jamal	Susie	Todd	Maribel	Grace
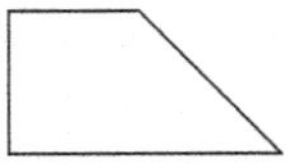					

Jamal has a shape that is a quadrilateral, but it is not a parallelogram.

Susie loves any shape that has more than 5 sides.

Todd likes shapes that have an infinite number of lines of symmetry.

Maribel likes shapes that have all right angles.

Grace likes quadrilaterals that have only 1 right angle.

Tuesday: Vocabulary Bingo

Put these words on your bingo board. Do not put them in the order that they are written. Scatter them in different places. Your teacher will call them out. Mark them on your board. Whoever gets 4 in a row horizontally, vertically or as a postage stamp (4 in a corner) first wins.

Words: multiple, composite, odd, even, prime, factor, ray, degree, digit, benchmark fraction, decompose, common denominator, equivalent.

Wednesday: Greater Than, Less Than, in Between

$\frac{1}{4}$ $\frac{2}{3}$ $\frac{9}{8}$

Name a fraction less than $\frac{1}{4}$	Name a fraction in between $\frac{2}{3}$ and $\frac{9}{8}$	Name a fraction less than $\frac{2}{3}$
Name a fraction greater than $\frac{9}{8}$	Name a fraction in between $\frac{1}{4}$ and $\frac{2}{3}$	Name a fraction greater than $\frac{2}{3}$

Thursday: British Number Talk

Pick a number from each circle. Make an addition problem. Then, decide how you are going to solve it. Write the problem under the way you solved it.

I can do it in my head.	I can do it with a model.	I can do it using a written strategy or algorithm.

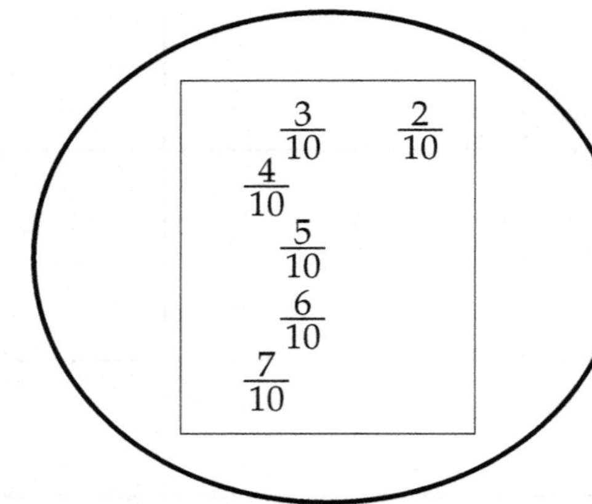

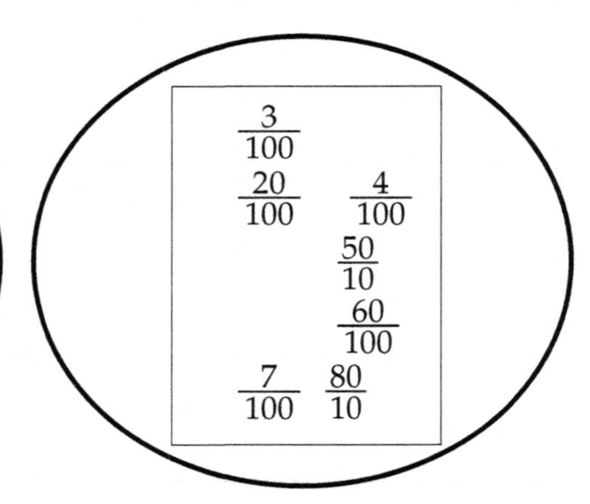

Friday: What's the Story? Here's the Model

Write a story that matches the tape/strip diagram. Write the answer.

20			
?	?	?	?

Story:

Week 33 Teacher Notes

Monday: Draw That!

Students have to reason about the shapes and draw them.

Tuesday: Vocabulary Tic Tac Toe

Students play rock, paper, scissors to see who goes first. Then take turns, picking a square, saying what it means, drawing or writing something about the word on the side and then marking the word with an x or an o. Whoever gets 3 in a row first wins.

Wednesday: Model It

Students model 3 different decimals. Write them. Describe them. Draw a number line and order them from least to greatest. Explain their reasoning.

Thursday: What's Missing?

Students fill in the missing numbers from the multiplication puzzle.

Friday: What's the Story? Here's the Graph

Students have to read the graph and then tell a story and ask and answer questions about it.

Week 33 Activities

Monday: Draw That!

Draw 1 shape that has 1 right angle.

Draw 1 shape that has 4 right angles.

Draw a shape that has 1 set of parallel sides.

Tuesday: Vocabulary Tic Tac Toe

Play rock, paper, scissors to see who goes first. Then take turns, picking a square, saying what it means, drawing or writing something about the word on the side and then marking the word with an x or an o. Whoever gets 3 in a row first wins.

gram	kilogram	milligram
kiloliter	liter	ml
m	cm	kilometer

ounces	pint	quart
yard	ft	pound
inch	gallon	customary measurement

Wednesday: Model It

Model 3 different decimals. Write them down. Describe them. Plot on the number line and order them from least to greatest. Explain your reasoning.

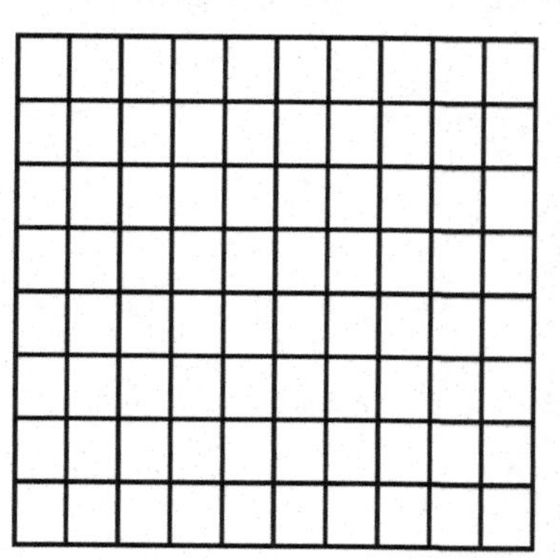 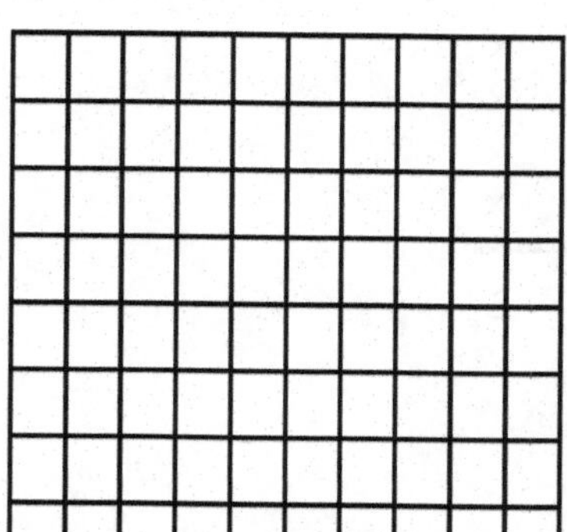 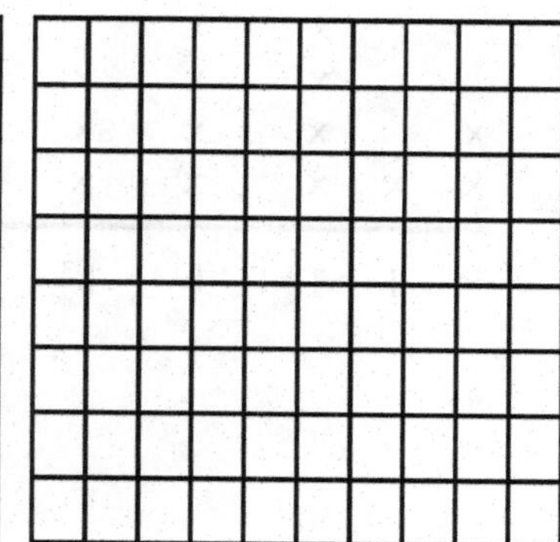

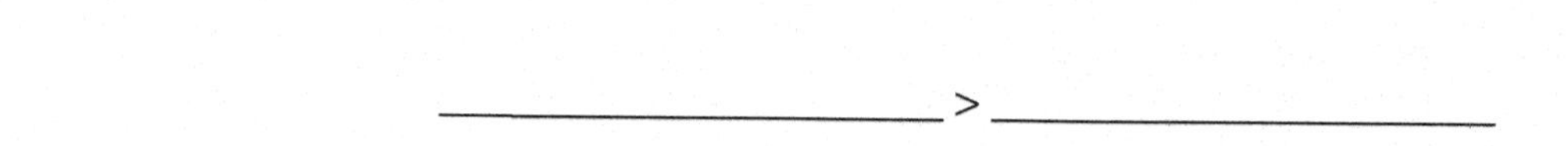

____________________ > ____________________

____________________ < ____________________

Thursday: What's Missing?

$$
\begin{array}{r}
1\ \square \\
\times\ \square\ 1 \\
\hline
12 \\
120 \\
\hline
\square\square\square
\end{array}
$$

Explain your thinking to a partner.

Be able to defend your thinking in the whole group discussion.

Friday: What's the Story? Here's the Graph

Here's the graph.

		x	x		x
		x	x	x	x
x		x	x	x	x
x	x	x	x	x	x
$\frac{1}{2}$	1	$1\frac{1}{8}$	$1\frac{1}{2}$	$1\frac{3}{4}$	$2\frac{3}{8}$

What's the story? Make up a story and tell it to your partner. Explain your thinking. Be ready to share your thinking to the whole group.

Week 34 Teacher Notes

Monday: 2 Arguments

Students read the 2 arguments and decide who they agree. They have to say why and prove their thinking with the decimal grids. They also have to plot the decimals on the number line.

Tuesday: Vocabulary Bingo

Students have to put these numbers on their bingo board. They should not put them in the order that they are written. They should scatter them in different places. The teacher will call them out. Students mark them on their board. . Whoever gets 4 in a row horizontally, vertically or as a postage stamp (4 in a corner) first wins. Call out the words using pictures and definitions.

Wednesday: What Doesn't Belong?

Students have to decide which numbers do not belong in each set and then explain their thinking.

Thursday: Find and Fix the Error

Students have to find and fix the error. They have to discuss their thinking with a math partner.

Friday: Model It

Students have to tell a story about the scenario and model and solve it.

Monday: 2 Arguments

Mike said that .4 is not equivalent to .40.
Sara said that it is equivalent.

Who do you agree with? Why?

Prove it in 2 ways.

Grid:

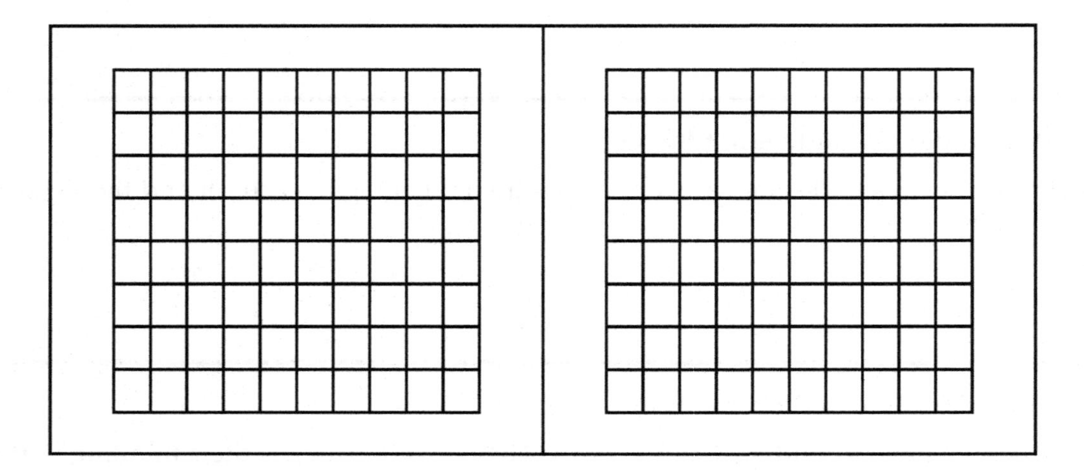

Prove it on a number line.

Tuesday: Vocabulary Bingo

Put these words on your bingo board. Do not put them in the order that they are written. Scatter them in different places. Your teacher will call them out. Mark them on your board. Whoever gets 4 in a row horizontally, vertically or as a postage stamp (4 in a corner) first wins.

Words: quadrilateral, trapezoid, square, rhombus, rectangular, circle, pentagon, polygon, parallelogram, triangle, parallel, vertex hexagon, figure, angle, perpendicular.

Wednesday: What Doesn't Belong?

Look at each set. Decide which item does not belong in each set. Discuss with your partner and the class.

A.

8	.8
.800	.80

B.

trapezoid	square
rectangle	rhombus

Thursday: Find and Fix the Error

Read the problem. Discuss Miguel's error. Solve it correctly.

Miguel put 2 as the answer. Explain what he did incorrectly. Discuss what he should do.

$\_\_\_\_\_ \div 5 = 10$

Explain your thinking to a partner.

Be able to defend your thinking in the whole group discussion.

Friday: Model It

The areas of two shapes are very similar, but the perimeters of those two shapes are very different. What might the shapes be?	

Week 35 Teacher Notes

Monday: Convince Me!

Students have to think about the problem and then convince their partner that it is true.

Tuesday: Talk and Draw

Students have to talk about the topic and draw pictures to explain their thinking.

Wednesday: 3 Truths and a Fib

Students have to look at the statements and pick the one that is false.

Thursday: Find and Fix the Error

Students have to look at the problem, find the error and then explain how to fix it and do it correctly.

Friday: Regular Word Problem

Students have to read and solve the problem.

Week 35 Activities

Monday: Convince Me!

Convince me that $4 \times 2 = 16 \div 2$

Tuesday: Talk and Draw

Tell your partner everything you know about lines. Draw some examples. Use some of the vocabulary: parallel, perpendicular, intersecting, straight, curved, points, line segments, rays, angles, right angles.

Wednesday: 3 Truths and a Fib

Read statements. Decide which one is false. Discuss.

450 is the same as 45 tens.
$\frac{2}{3}$ is equivalent to $\frac{4}{6}$.
305 is the same as 3 hundreds and 5 tens.
$\frac{4}{4}$ is the same as 1.

Thursday: Find and Fix the Error

Read the problem. Figure out Kevin's error. Solve the correct way. Discuss.
Kevin solved the problem this way:

$$\frac{3}{10} + \frac{50}{100} = \frac{53}{100}$$

Why is it wrong?

What did he do wrong?

How can we fix it?

Friday: Regular Word Problem

Read the problem and solve. Draw a model. Discuss your thinking.

At the bicycle shop there were bicycles and tricycles. There were 20 wheels. How many bicycles and tricycles could there have been?

Week 36 Teacher Notes

Monday: Subtraction Puzzle

Students have to reason about the puzzle and fill in the missing numbers.

Tuesday: Vocabulary Tic Tac Toe

Students play rock, paper, scissors to see who goes first. Then they take turns, picking a square, saying what it means, drawing or writing something about the word on the side and then marking the word with an x or an o. Whoever gets 3 in a row first wins.

Wednesday: Number Bond It!

Students have to break apart the fractions in different ways.

Thursday: British Number Talk

Students pick a number from each circle. Tell them to make an addition problem. They should write the problem under the way they solved it. For example, "I can do that in my head. I added 165 and 29. I made the problem 164 and 30 and I got 194."

Friday: Regular Word Problem

Students have to read and solve the problem.

Monday: Subtraction Puzzle

Fill in the blank spaces.

```
  2 □ 9 2
–   3 □ 9
---------
  □ 8 2 3
```

1. Think about it.
2. Share your thinking with a friend. Defend your answer.
3. Share your thinking with the group.

Tuesday: Vocabulary Tic Tac Toe

Play rock, paper, scissors to see who goes first. Then take turns, picking a square, saying what it means, drawing or writing something about the word on the side and then marking the word with an x or an o. Whoever gets 3 in a row first wins.

acute	ray	right angle
straight angle	line	360 degrees
obtuse angle	endpoint	degree

multiple	quotient	sum
square number	divisor	difference
factor	dividend	product

Wednesday: Number Bond It!

Show how to decompose $\frac{7}{8}$ in 3 different ways!

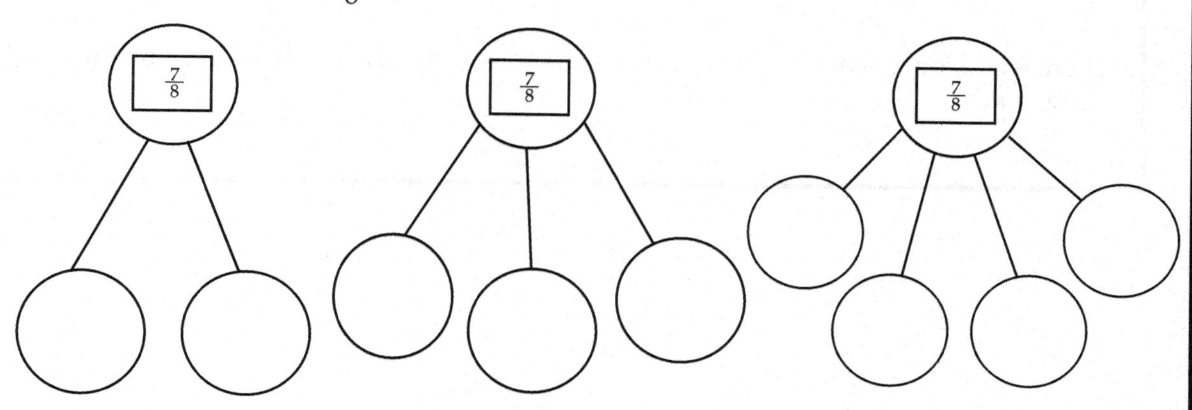

Thursday: British Number Talk

Pick a number from each circle. Make a fraction subtraction problem. Use fractions with like denominators. Write the problem under the way you solved it.

I can do it in my head.	I can do it with a model.	I can do it using a written strategy or algorithm.

A.

$\frac{1}{2}$ $\frac{3}{3}$ $\frac{1}{3}$ $\frac{2}{3}$ $\frac{4}{5}$

$\frac{1}{4}$ $\frac{3}{4}$ $\frac{1}{5}$ $\frac{2}{5}$ $\frac{3}{5}$

$\frac{1}{6}$ $\frac{2}{6}$ $\frac{5}{6}$ $\frac{3}{6}$

$\frac{4}{6}$ $\frac{6}{6}$ $\frac{2}{2}$ $\frac{2}{4}$

$\frac{1}{8}$ $\frac{2}{8}$ $\frac{3}{8}$ $\frac{5}{8}$

B.

$\frac{1}{2}$ $\frac{3}{3}$ $\frac{1}{3}$ $\frac{2}{3}$ $\frac{4}{5}$

$\frac{1}{4}$ $\frac{3}{4}$ $\frac{1}{5}$ $\frac{2}{5}$ $\frac{3}{5}$

$\frac{1}{6}$ $\frac{2}{6}$ $\frac{5}{6}$ $\frac{3}{6}$

$\frac{4}{6}$ $\frac{6}{6}$ $\frac{2}{2}$ $\frac{2}{4}$

$\frac{1}{8}$ $\frac{2}{8}$ $\frac{3}{8}$ $\frac{5}{8}$

Friday: Regular Word Problem

Read the problem and solve it.

There was a fence that had an area of 48 square feet. What could have been the dimensions of the fence?

Week 37 Teacher Notes

Monday: Always, Sometimes, Never

Students have to read the statement and decide if it is always true, sometimes true or never true.

Tuesday: Vocabulary Bingo

Students put the words in the box in a different order than they appear. They play bingo. When the teacher calls a word, they cover it. Whoever gets 4 in a row vertically, horizontally or 4 corners wins.

Wednesday: Guess My Number

Students have to use the clues to find the mystery fraction.

Thursday: British Number Talk

Students pick a number from each circle. They make an addition problem. They write the problem under the way they solved it. For example, "I can do that in my head. I added 165 and 29. I made the problem 164 and 30 and I got 194."

Friday: Fill in the Problem!

Students have to read the problem and fill in their own numbers and then solve it.

Week 37 Activities

Monday: Always, Sometimes, Never

Read the statement. Decide if this statement is always, sometimes or never true. Discuss your thinking.

Numbers ending in 4 are multiples of 4.

1. Think about it.
2. Share your thinking with a friend. Defend your answer.
3. Share your thinking with the group.

Tuesday: Vocabulary Bingo

Put these words on your bingo board. Do not put them in the order that they are written. Scatter them in different places. Your teacher will call them out. Mark them on your board. Whoever gets 4 in a row horizontally, vertically or as a postage stamp (4 in a corner) first wins.

Words: decimal, hundredths, tenths, twelfths, halves, decimal point, numerator, fraction, denominator, whole, equivalent, mixed number, hundreds, thousands, tens, ones.

Wednesday: Guess My Number

Read the clues. Decide which one is the mystery fraction.

I am a fraction that is greater than $\frac{5}{4}$. I am less than $2\frac{1}{2}$. My numerator is not odd. My denominator is not even. Who am I?	$1\frac{1}{4}$ $\frac{4}{2}$ $2\frac{2}{7}$ $2\frac{4}{5}$ $1\frac{4}{6}$

Thursday: British Number Talk

Pick a number from each circle. Make an addition problem. Write the problem under the way you solved it.

I can do it in my head.	I can do it with a model.	I can do it using a written strategy or algorithm.

A.

Make up your own numbers.

B.

Make up your own numbers.

Friday: Fill in the Problem!

Read the problem. Fill in the blanks. Model the story and solve.

Mike had ______ marbles. He put ______ in a box. He filled ______ boxes. He had _______ left over.

Draw a model of your story.

Write an equation that goes with the story.

Week 38 Teacher Notes

Monday: Pattern /Skip Counting

Students have to look at the pattern and figure out what comes next. They should be ready to discuss their thinking with a math partner.

Tuesday: Vocabulary Bingo

Students play bingo with the designated words. They put these numbers on your bingo board. They do not put them in the order that they are written. They should scatter them in different places. The teacher will call out the words and student will mark them on their board. Whoever gets 4 in a row horizontally, vertically or as a postage stamp (4 in a corner) first wins.

Wednesday: Place Value Puzzle

Students have to reason about the numbers and make a problem based on the directions.

Thursday: British Number Talk

Students pick a number from each circle. They decide how they will solve it and write that expression under the title.

Friday: Make Your Own Problem!

Students write their own fraction word problem.

Week 38 Activities

Monday: Pattern/Skip Counting

Make a pattern starting with the number 35 where you add 7 and then subtract 2. Make sure to complete 6 terms.

35, _____, _____, _____, _____. _____, _____

Now make a pattern that follows the same rule as the pattern above, but start with a number with 4 digits. Complete 6 terms.

_______, ________, ________, ________, _______, ________

Tuesday: Vocabulary Bingo

Put these words on your bingo board. Do not put them in the order that they are written. Scatter them in different places. Your teacher will call them out. Mark them on your board. Whoever gets 4 in a row horizontally, vertically or as a postage stamp (4 in a corner) first wins.

Words: elapsed time, half past, quarter til, quarter after, hour, minute hand, hour hand, decimal, tenths, hundredths, decimal point, equivalent. whole number, fraction, millions, ten thousands.

Wednesday: Place Value Puzzle

Think about this problem with a partner. Solve. Be ready to discuss in the whole group.

Jan picked the numbers 2, 3, 5, 8, 9, 0.

How should Jan arrange the numbers to get the largest sum? Write the numbers in the boxes and then find the sum.

Thursday: British Number Talk

Pick a number from each circle. Make a subtraction problem. Write the problem under the way you solved it.

I can do it in my head.	I can do it with a model.	I can do it using a written strategy or algorithm.

A. Write your own 4 digit numbers.

B. Write your own 4 digit numbers.

Friday: Make Your Own Problem!

Write a word problem about fractions.

Week 39 Teacher Notes

Monday: Why Is It Not?

Students have to explain why it is not the given answer.

Tuesday: What Doesn't Belong?

Students have to look at the set and find the one that doesn't belong and explain their thinking.

Wednesday: Place Value Puzzle

Students have to look at the information and reason about the numbers. They should explain their thinking.

Thursday: British Number Talk

Students pick a number from each circle. They make an addition problem. They write the problem under the way they solved it. For example, "I can do that in my head. I added 165 and 29. I made the problem 164 and 30 and I got 194."

Friday: Make Your Own Problem!

Students have to write a story where the answer is 4/10ths.

Week 39 Activities

Monday: Why Is It Not?

A. $9 \times \_\_\_ = 81$

Why is it not 89?

B. $7 = \frac{?}{11}$

Why is it not 18?

Tuesday: What Doesn't Belong?

Look at each set. Decide which word does not belong in each set. Discuss with your partner and the class.

A. What doesn't belong?

cm	km
mm	kl

B. What doesn't belong?

pints	liter
quarts	grams

Wednesday: Place Value Puzzle

Think about this problem with a partner. Solve it. Be ready to discuss in the whole group.

Jan picked the numbers 2, 3, 5, 8, 7, 1.

How should Jan arrange the numbers to get the largest difference? Write the numbers in the boxes and then find the difference.

	☐	☐	☐
–	☐	☐	☐

Thursday: British Number Talk

Pick a number from each circle. Make a multiplication problem. Write the problem under the way you solved it.

I can do it in my head.	I can do it with a model.	I can do it using a written strategy or algorithm.

A. Write your own 2 digit numbers.

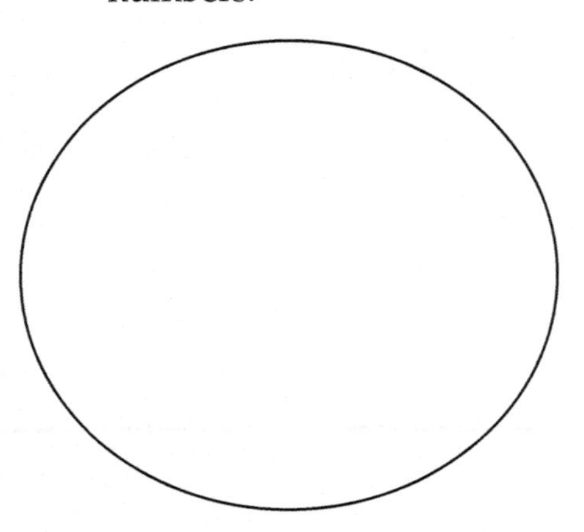

B. Write your own 2 digit numbers.

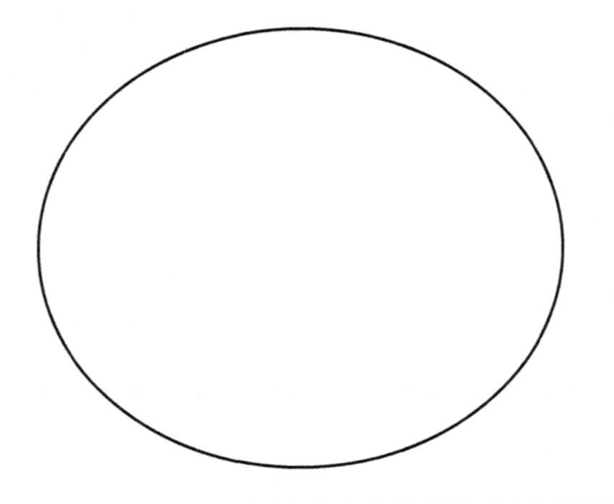

Friday: Make Your Own Problem!

The answer is $\frac{4}{10}$. Write a word problem.

Story:

Week 40 Teacher Notes

Monday: Missing Number

Students fill in the missing number equations to make them true.

Tuesday: Vocabulary Tic Tac Toe

Students write down words they choose in the tic tac toe board. They play with a partner.

Students play rock, paper, scissors to see who goes first. Then they take turns, picking a square, saying what it means, drawing or writing something about the word on the side and then marking the word with an x or an o. Whoever gets 3 in a row first wins.

Wednesday: Number of the Day

Students choose their own number and fill in the chart to represent the number.

Thursday: British Number Talk

Students put their own numbers in the circles and then pick a number from each circle. They do calculations with those numbers. Students then write the problem under the way they solved it.

Friday: Make Your Own Problem!

Students write their own word problems.

Week 40 Activities

Monday: Missing Number

Fill in the missing numbers to make the equations true.

70 = ___ ÷ ___

100 = ___ × ___

75 = ___ + ___ + ___

1025 = ___ − ___

Tuesday: Vocabulary Tic Tac Toe

Start by choosing words to write in the mats. Play rock, paper, scissors to see who goes first. Then take turns, picking a square, saying what it means, drawing or writing something about the word on the side and then marking the word with an x or an o. Whoever gets 3 in a row first wins.

Wednesday: Number of the Day

________ Write a 4-digit number

Word form	Expanded form	100 more than _____ is ______ 100 less than ____ is _______.
____ is greater than_____ ____ is less than _______	____ + ____ = _____	____ − ____ = ______

Plot it on the number line. Also put 2 numbers that are larger than the number and 2 numbers that are smaller than the number.

Thursday: British Number Talk

Pick a number from each circle. Make a subtraction problem. Write the problem under the way you solved it.

I can do it in my head.	I can do it with a model.	I can do it using a written strategy or algorithm.

A. Make up your own numbers. Put 2, 3 or 4 digit whole numbers.

B. Put your own single digit numbers in the circle.

Friday: Make Your Own Problem!

Write your own multiplication word problem. Use 2-digit numbers.

Story:

Model:

Equation:

Bonus Template 1:

Week:___

Monday: True or False?

Make your own true or false statements. Explain your thinking to a partner.

Equation	True or False?

Tuesday: Vocabulary Tic Tac Toe

Make up your own math vocabulary tic tac toe. Play with your partner.

Play rock, paper, scissors to see who goes first. Then take turns, picking a square, saying what it means, drawing or writing something about the word on the side and then marking the word with an x or an o. Whoever gets 3 in a row first wins.

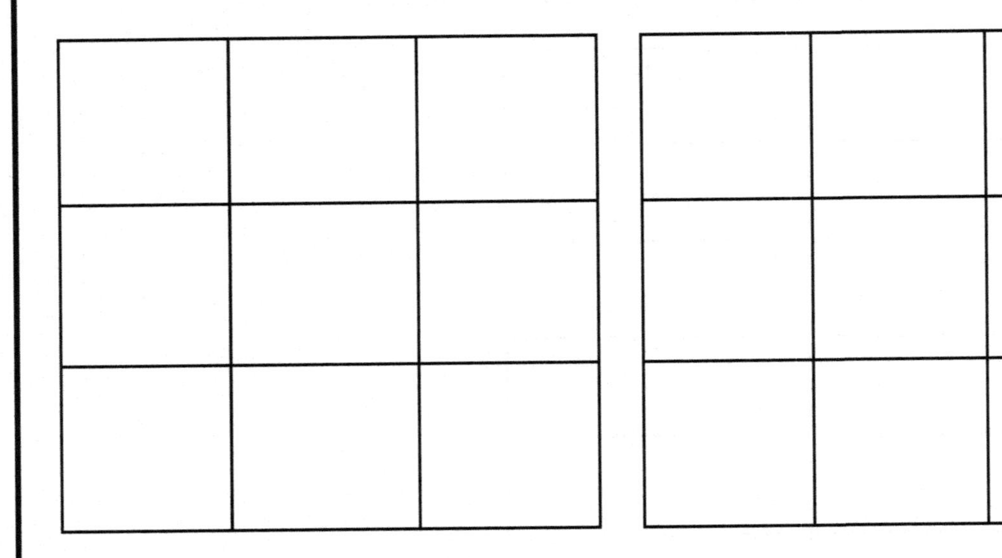

Wednesday: Fraction of the Day

Choose your own fraction of the day.

Fraction in word form.	Show an addition sentence that makes your number.
Make a subtraction sentence that makes your number.	Write 3 equivalent fractions.

Thursday: Number Talk

Write numbers in the circles. Pick a number from each circle. Make different problems. Use whatever operation you choose. Solve it. Write the problem under the way you solved it.

Can I do it in my head?	Do I need a model?	Am I going to write down and calculate the numbers?

A. Make up your own numbers.

B. Make up your own numbers.

Friday: Regular Word Problem

Write and answer your own multi-step word problem.

Bonus Template 2:

Week:___
Monday:
Tuesday:
Wednesday:
Thursday:
Friday:

Answer Key

Week 1

Monday: What Doesn't Belong?

A. 5 + 3 + 4.

B. 50 – 33.

Tuesday: Vocabulary Match

- Product – the answer to a multiplication problem.
- Sum – The result of adding numbers or quantities together.
- Difference – The amount left over when one number is being subtracted from another number.
- Quotient – the answer to a division problem.
- Factor – numbers that are multiplied together to find a product.

Wednesday: Convince Me!

Answers vary.

Thursday: Number Talk

Answers vary. Students might say that they made 128 into 130 by taking 2 from 9 and then they added 7 and 76 to get 83 and added 83 and 130 to get 213.

Friday: What's the Question? (3 Read Protocol)

Answers vary. For example: How many marbles does he have altogether?

Week 2

Monday: Magic Square

Magic number is 15.

Tuesday: Vocabulary Tic Tac Toe

Answers vary.

Wednesday: Number Line It!

Answers vary. $\frac{1}{4}$, and $\frac{2}{8}$ sit in the same place and then $\frac{1}{2}$, $\frac{4}{4}$ and then $\frac{5}{4}$.

0 $\quad \frac{1}{4}$ $\frac{2}{8}$ $\quad \frac{1}{2}$ $\quad \frac{4}{4}$ $\quad \frac{5}{4}$ $\quad$ 1

Thursday: British Number Talk

Answers vary. For example, 218 – 118 is 100.

Friday: Make Your Own Problem!

Answers vary. For example: The rug is 9 feet long and 2 feet wide. What is its area?

Week 3

Monday: Always, Sometimes, Never

This is sometimes true. Example: all parallelograms are quadrilaterals, but there are other types of quadrilaterals such as trapezoids, rectangles, squares, and rhombi.

Tuesday: Frayer Model

Quadrilaterals are 4 sided figures. An example is a square. A nonexample would be a circle.

Wednesday: Number of the Day

One thousand one hundred ninety nine. Answers vary for addition and subtraction sentences. For example, 1,000 + 199 or 2,299 -1,100. It is composite. It is odd. It rounds to 1,200.

Thursday: Number Strings

Answers vary. Students should discuss how you can think about numbers in relationship to each other. For example, if 25 + 25 is 50 then 25 + 27 is 52.

Friday: Model It

Students may draw 3 hundreds, 9 tens, and 9 ones and add 1 hundred, 4 tens, and 4 ones and show the regrouping of the hundreds and ones to get to the sum of 543.

Week 4

Monday: Magic Square

2	7	6
9	5	1
4	3	8

Tuesday: It Is/It Isn't

Answers vary. For example: It is a 4-sided figure. It does not have curved sides.

Wednesday: Greater Than, Less Than, in Between

Answers vary.

Name a fraction less than $\frac{1}{3}$ $\frac{1}{6}$	Name a fraction in between $\frac{1}{3}$ and $\frac{6}{6}$ $\frac{4}{6}$	Name a fraction that is less than $\frac{1}{6}$ $\frac{1}{12}$
Name a fraction greater than $\frac{1}{3}$ $\frac{2}{3}$	Name a fraction greater than $\frac{6}{6}$ $\frac{7}{6}$	Name a fraction in between $\frac{1}{6}$ and $\frac{6}{6}$ $\frac{5}{6}$

Thursday: Number Talk

Answers vary. Add 1 to each number so the problem becomes. 1,405 – 1,300 = 105.

Friday: Picture That!

Answers vary.

Week 5

Monday: 2 Arguments

Maria is correct.

Tuesday: Vocabulary Match

Match the vocabulary with the example.

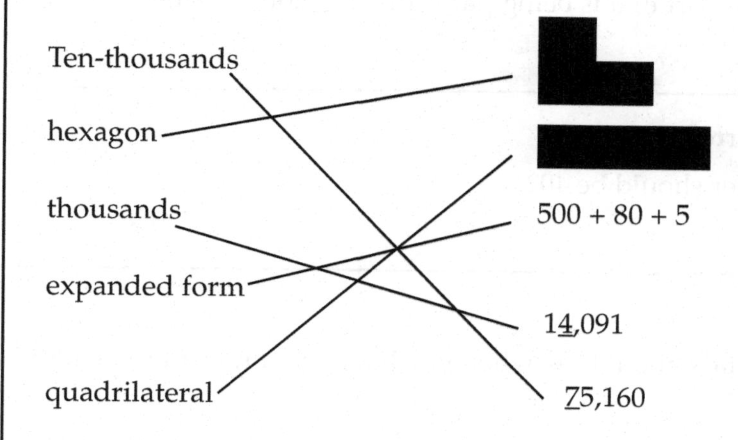

Wednesday: Money Combinations

Answers vary. For example: 7 quarters and 14 pennies.

Thursday: Number Talk

Answers vary. For example, 250 - 124.

Friday: Make Your Own Problem!

Answers vary. For example: Toy store A had 245 marbles. Toy store B had 345 marbles. Toy store B had 100 more marbles than toy store A.

(Models vary).

245 + 100 = 345

Week 6

Monday: It Is/It Isn't

Answers vary. For example: In standard form the number is 20,346. It does not contain any thousands.

Tuesday: 1-Minute Essay

Answers vary. For example: A divisor is the number that tells how many parts to partition the number into. Dividend is the part that is being partitioned. Quotient is the answer.

Wednesday: Find and Fix the Error

John did not regroup. The answer should be 402.

Thursday: Number Strings

Answers vary. Students should talk about how when you have a 9 you want to make a 10 to have an easier problem.

Friday: Equation Match

Answer is C.

Week 7

Monday: 3 Truths and a Fib

A car is about 40 yards long is false.

Tuesday: Vocabulary Brainstorm

Answers vary. Multiplication is the opposite of division.

Wednesday: Pattern/Skip Counting

1. $\frac{1}{2}, \frac{3}{4}, \frac{4}{4}, 1\frac{1}{4}, 1\frac{1}{2}, 1\frac{3}{4}, 2, 2\frac{1}{4}, 2\frac{1}{2}$
2. $\frac{2}{3}, 1\frac{1}{3}, 2, 2\frac{2}{3}$ and $3\frac{1}{3}$
3. Make your own: Answers vary.

Thursday: Number Talk Puzzle

339 + 787.

Friday: Make Your Own Problem!

Answers vary. The bakery had 11 cupcakes. They put 3 in a box. How many boxes did they use? How many cupcakes did they have left over?

Week 8

Monday: True or False?

Some of these are hexagons.

Tuesday: Vocabulary Tic Tac Toe

Answers vary.

Wednesday: Number Bond It!

Answers vary. For example: 3,942 = 3,900 + 42

Thursday: Number Talk

Answers vary. For example: 200 + 357.

Friday: Model It

Answers vary.

For example: Joe had 2 marbles. His brother had 2 times as many as he did. How many did they have altogether?

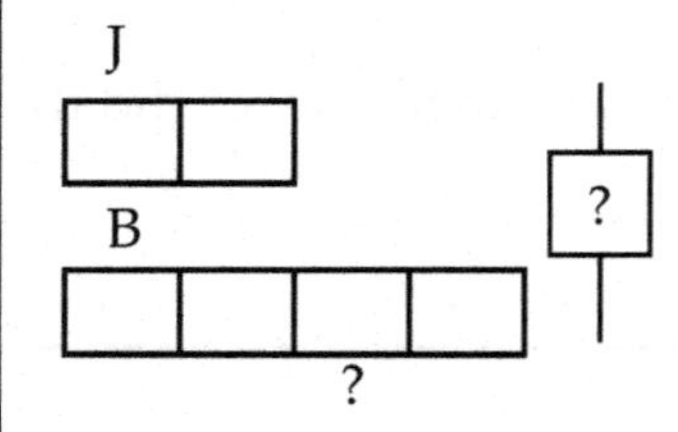

Week 9

Monday: Convince Me!

Answers vary.

Tuesday: Vocabulary Match

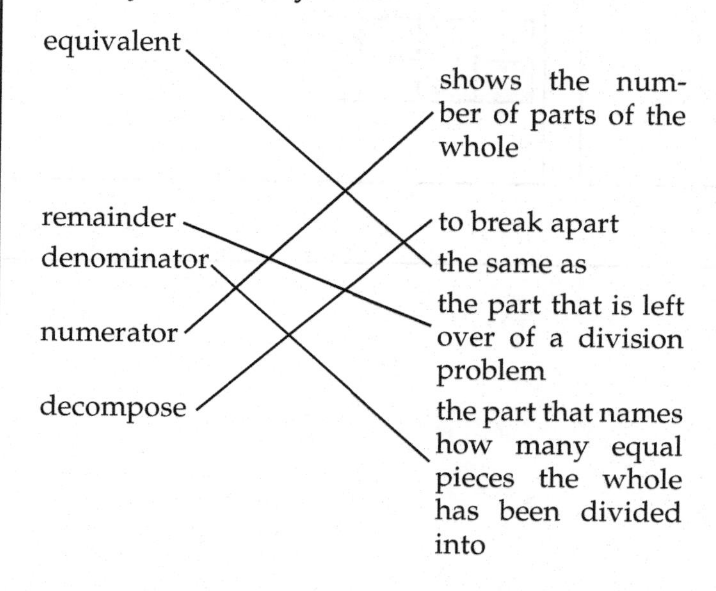

Wednesday: Fraction of the Day

$\frac{5}{8}$

Word form: Five-eighths	How much more is needed to get to 1 whole? $\frac{3}{8}$	Draw a model:
Plot it on the number line: Add a fraction that is greater than this fraction. Add a fraction that is less than this fraction. Plot an equivalent fraction on the number line. $\frac{2}{8}$ $\frac{5}{8}$ $\frac{10}{16}$ $\frac{8}{8}$		$\frac{5}{8} = \frac{10}{16}$ $\frac{5}{8} > \frac{4}{8}$ $\frac{3}{8} < \frac{5}{8}$

Thursday: Number String

Answers vary. Students should discuss making tens. For example, 142 - 27 = 145 - 30

Friday: Model It

$\frac{3}{4}$ $\frac{6}{8}$

Week 10

Monday: Reasoning Matrices

	Pepperoni	Cheese	Chicken	Mushroom	Veggies	Pepperoni and Pineapple
Jenny		x				
Jamal	x					
Miguel			x			
Kelly						x
Maria				x		
Grace					x	

Tuesday: 1-Minute Essay

Answers vary. For example: Polygons are closed figures with straight sides.

Wednesday: Number of the Day

The 7 in the tens place is ten times the amount of the 7 in the ones place.

7,277

Word form: Seven thousand, two hundred seventy seven	10 more: 7,287	10 less: 7,267
expanded form: 7,000 + 200 + 70 + 7	7,000 + 277 = 7,277	8,277 – 1,000 = 7,277
100 more: 7,377	How many more to 10,000? 2,723	odd or even? odd
1,000 more: 8,277	100 less: 7,177	7,247 – 247 = 7,000

Thursday: Number Talk

Answers vary.

Friday: Make Your Own Problem!

Answers vary. Marcia drank 500 ml of water in the morning and 500 ml in the afternoon. In the evening she drank 1,000 ml. How many liters did she drink altogether?

Week 11

Monday: Venn Diagram

Answers vary. Students could have rhombi, trapezoids and hexagons for polygons. They could have rectangles and squares and right-angled triangles for polygons with right angles.

Tuesday: Vocabulary Bingo

Answers vary.

Wednesday: How Many More to

A.	B.
1. Start with 25 get to 100. 75	1. Start with $\frac{1}{5}$ get to 1. $\frac{4}{5}$
2. Start with 179 get to 200. 21	2. Start with $\frac{2}{12}$ get to $\frac{4}{6}$. $\frac{6}{12}$
3. Start with 589 get to 1000. 411	3. Start with $\frac{2}{3}$ get to 1. $\frac{1}{3}$

Thursday: What's Missing?

A. 8 ÷ **2** = 4.
B. **1000** ÷ 10 = 100.
C. 25 ÷ **5** = 5.
D. ___ ÷ ____ = ____ Answers vary.

Friday: What's the Question? (3 Read Protocol)

Answers vary. For example: How much did Joe have? How much did they eat altogether?

Week 12

Monday: Legs and Feet

A. 6.
B. Answers vary. 2 cows and 2 chickens.
C. Answers vary. 5 chickens and 1 cow.
D. Answers vary. 6 chickens and 2 cows.
E. Answers vary. 12 legs. 24 legs could be 2 chickens, 2 cows and 2 crickets.
F. Answers vary. 4 crickets, a chicken and a cow.

Tuesday: Vocabulary Brainstorm

Answers vary. For example: Length measures how long something is.

Wednesday: 3 Truths and a Fib

$\frac{2}{4} = \frac{5}{8}$ is a fib.

Thursday: Number Talk

You want students to think about counting up or adding 172 to each number to get a more friendly number. 10,172 – 2,000 is an easier problem.

Friday: Make Your Own Problem!

Answers vary. For example: The bakery had 11 cookies. They put 2 in a box. They used 5 boxes and they had 1 cookie leftover.

Week 13

Monday: Patterns/Skip Counting

Part A. C. (9, 20, 42, 86).
Part B. Answers vary. For example: 6, 14, 30, 62.

Tuesday: What Doesn't Belong?

A. factor.

B. difference

Wednesday: Rounding

1. Answers vary. For example, 492, 540 and 501.
2. Answers vary. For example, 6,789; 6,999; 7,401.
3. Answers vary. 17,101; 17,057; 17,099.

Thursday: Number Talk

Answers vary. Students can talk about breaking apart the problem into 30×7 plus 5×7.

Friday: What's the Question? (3 Read Protocol)

Answers vary. How many gold rings did they have? How many silver rings did they have? How many rings did they have altogether? How many more silver rings are there than gold rings?

Week 14

Monday: Convince Me!

Answers vary.

Tuesday: Frayer Model

Answers vary. For example: definition: break apart; examples 789 = 700 + 80 + 9; a trapezoid can be decomposed into 3 triangles. Non-example: multiply.

Wednesday: Guess My Number

A. 72.

B. 288.

Thursday: Number Talk

Answers vary. For example: break apart 415 into 350 and 63 with a remainder of 2 so you get partial quotients of 50 and 9 which is 59 with a remainder of 2. This can be represented as $59\frac{2}{7}$.

Friday: Regular Word Problem

190 + 240 + 225 = \$655. They sold 131 pounds of chocolate. They made \$655.

Week 15

Monday: Number Line It!

Here is the answer and possible fractions students could add.

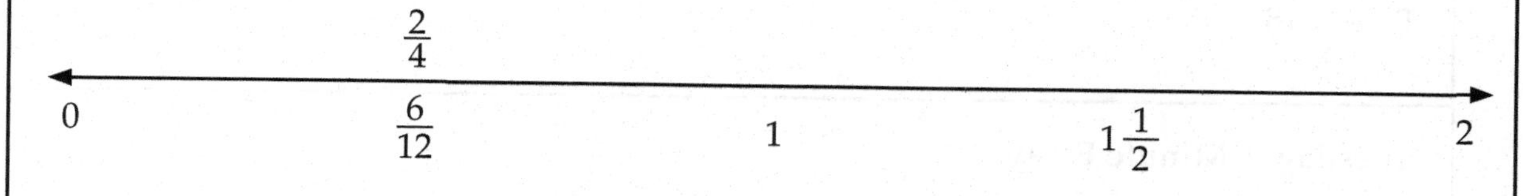

Tuesday: Vocabulary Tic Tac Toe

Answers vary.

Wednesday: Venn Diagram

Answers vary. For example: Fractions: $\frac{1}{2}$, $\frac{3}{4}$, $\frac{5}{8}$. Fractions that are equivalent to $\frac{1}{2}$: $\frac{4}{8}$, $\frac{6}{12}$.

Thursday: British Number Talk

Answers vary. For example: 2×15 is 30.

Friday: Time Problem

Answers vary. He could have left at 4 and come back at 5:30.

Week 16

Monday: What Doesn't Belong?

A. 100 – 45.

B. 40 – 18.

Tuesday: 1-Minute Essay

Answers vary. For example: Multiplication can be a fast way to add.

Wednesday: Fraction of the Day

$$\frac{7}{12}$$

Word form seven twelfths	Rectangle model
Equivalent fraction $\frac{14}{24}$	Draw a line and cut it into 12 parts and put a point at the 7th tick 1 2 3 4 5 6 $\frac{7}{12}$ 8 9 10 11 12 Answers vary for fractions that are greater and less than $\frac{7}{12}$ $\frac{9}{12}$ is greater than $\frac{7}{12}$ and $\frac{2}{12}$ is less than $\frac{7}{12}$.

Thursday: Number Talk

Answers vary. For example: 2×457 is 2×450 plus 2×7. This is 900 + 14 which is 914.

Friday: Sort That!

Answers may vary. For example:

Group 1: $\frac{1}{4}, \frac{1}{8}, \frac{1}{3}$ (Less than one half)

Group 2: $\frac{4}{10}, \frac{6}{10}$ (Close to $\frac{1}{2}$)

Group 3: $\frac{3}{4}, \frac{4}{5}$ (Close to 1)

Week 17

Monday: Input/Output Table

What's the rule? Multiply by 7		What's the rule? Divide by 5		Make your own.
A.		B.		C.
In	Out	In	Out	
4	28	20	4	
5	**35**	40	8	
6	42	15	**3**	Answers vary.
7	49	5	**1**	
8	**56**	**0**	0	
0	**0**	**10**	2	

Tuesday: Vocabulary Tic Tac Toe

Answers vary.

Wednesday: What Doesn't Belong?

A. $\frac{6}{8}$.

B. $\frac{4}{4}$.

Thursday: Number Strings

Answers vary. Students should discuss the relationships between 10s and 5s and 10s and 20s.

Friday: Fill in the Problem!

Answers vary. For example: Mike had 12 marbles. He put 2 in each box. How many boxes did he use? He used 6 boxes. He did not have any left over.

Week 18

Monday: Missing Number

A. $35 \div \mathbf{5} = 7$.

B. $5 \times \mathbf{2 \times 10} = 100$. Answers vary.

C. $35 + \mathbf{35 + 20} = 90$. Answers vary.

D. $55 = 100 - \mathbf{45}$.

Tuesday: What Doesn't Belong?

A. ml.

B. ounces.

Wednesday: Bingo

Answers vary.

Thursday: Number Talk

Answers vary.

Friday: Model It

Models vary. For example: Students could use a tape diagram, a number line or a sketch.

Week 19

Monday: Break It Up!

3×7

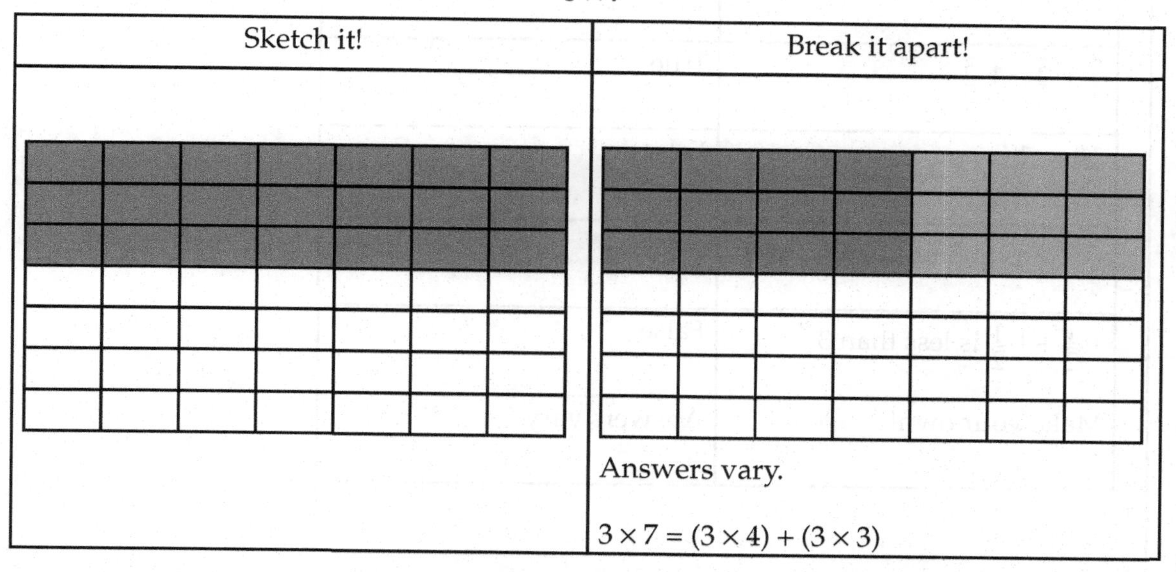

Tuesday: 1-Minute Essay

Answers vary. For example: Division is about sharing things out. When doing division you could be looking for the amount of groups or the amount in each group.

Wednesday: How Many More to

Students do this orally with a partner.

Thursday: British Number Talk

Answers vary. For example: 3,785 + 99. Make the 99 a 100 and then add 3,785 + 100 and take away 1.

Friday: What's the Story? (Here's the Model)

Answers vary. Sue had 4 boxes with 3 marbles in each box. Her sister had 6 boxes with 3 marbles in each box. How many did each girl have? How many marbles did they have altogether?

Week 20

Monday: True or False?

	True or False?
$\frac{5}{5}+\frac{3}{5}=1\frac{3}{5}$	True
$\frac{16}{8}=\frac{1}{2}$	False
$\frac{3}{4}+\frac{3}{4}$ is greater than $\frac{6}{8}$	False
$1\frac{1}{2}+1\frac{1}{2}$ is less than 3	False
Make your own!	Answers vary.

Tuesday: Vocabulary Brainstorm

Answers vary. For example: Equivalent means the same as.

$\frac{1}{2}=\frac{4}{8}$

Wednesday: Find and Fix the Error

He multiplied the top and bottom of the fraction by 4. He should have only multiplied the top by 4.

Thursday: Number Talk

Answers vary. For example: $\frac{10}{10}+\frac{2}{10}$ or $1+\frac{2}{10}$

Friday: What's the Story? (Here's the Model)

Answers vary. Mike had 20 marbles. He had 4 times as many as his brother. How many did his brother have?

Week 21

Monday: True or False?

A. True.

B. True.

C. Answers vary.

Tuesday: Vocabulary Bingo

Answers vary.

Wednesday: Number Bond It!

Answers vary. For example: $\frac{1}{2}$ gallon + $\frac{1}{2}$ gallon. 1 quart + 1 quart + 1 quart + 1 quart.

Thursday: British Number Talk

Answers vary. For example: 10×10 is 100. I just knew that. I did it in my head.

Friday: What's the Story?

Answers vary. For example: The bakery had 14 cookies. They packaged 5 cookies in a box. Each box had to have only 5 cookies. How many boxes did they use? How many cookies were left over?

Week 22

Monday: Missing Numbers

$\frac{7}{2} - \frac{2}{2} = \frac{5}{2}$. You only subtract the numerators. The denominators stay the same.

Tuesday: Frayer Model

Answers vary.

Remainder

Definition	Examples
A remainder is the part left over from a division problem after everything has been shared out or partitioned equally.	Sue had 10 cookies. She shared them equally among 3 friends. Was there any left over? How much?
Give a Picture Example	**Non-examples**
OOO | OOO | OOO | O $10 \div 3 = 3\frac{1}{3}$	

Wednesday: Number of the Day

82,091

Word form eighty two thousand ninety one	Expanded form 80,000 + 2,000 + 90 + 1	100 more is 82,191 100 less is 81,991
90,000 is greater than 82,091 72,500 is less than 82,091	82,000 + 91 = 82,091	90,000 – 7,909 = 82,091

0 25,000 50,000 82,091 92,000 100,000

Thursday: Number Talk

Answers vary. For example: $(300 \times 4) + (40 \times 4) + (6 \times 4)$ which is equal to $1{,}200 + 160 + 24 = 1{,}384$

Friday: What's the Question? Here's the Graph

Answers vary. For example: In Maribel's garden she found several plants. 2 were $\frac{1}{2}$ inch. 4 were 1 inch. 5 were $1\frac{1}{2}$ inches. 3 were 2 inches. 6 were $2\frac{1}{2}$ inches long.

Week 23

Monday: Magic Square

1	**15**	14	4
12	6	**7**	9
8	**10**	**11**	**5**
13	3	**2**	16

Tuesday: Frayer Model

Area

Definition	Examples
The amount of space an object covers.	We measure the area of rugs.
Give a Picture Example	**Non-examples**
3 ft 4 ft The area is 12 square feet	Perimeter

Wednesday: Missing Number

A. 72, 84, 96, 108, 120, 132, 144.
B. 320, 160, 80, 40, 20, 10, 5.
C. Make your own pattern: Answers vary.

Thursday: Number Talk

Answers vary. For example: 62/9

Think about 54 and then the remainder . . .

It would be 6 8/9.

Friday: Model It

t

15	15

f

15

?

$2 \times 15 = 30$

There were 15 fish and 30 turtles. There were 45 animals altogether.

Week 24

Monday: Input/Output Table

Answers vary. For example:

In	Out
1	11
2	22
3	33
4	44

Tuesday: 1-Minute Essay

Answers vary. For example: Mixed numbers are whole numbers and fractions together.

Wednesday: Bingo

Answers vary.

Thursday: British Number Talk

Answers vary. For example: $\frac{28}{4} = 7$.

Friday: Model It

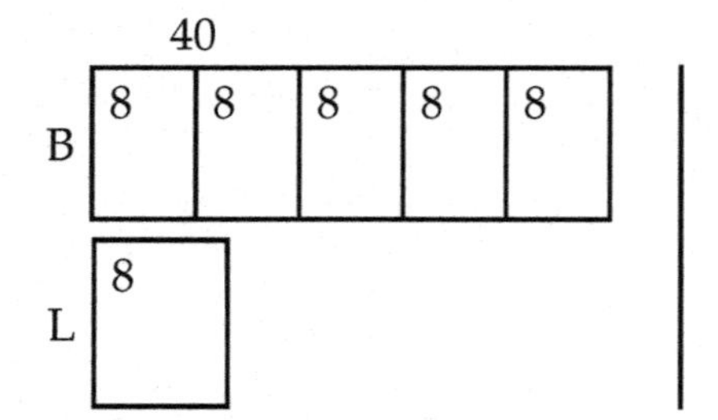

There were 8 ladybugs. There were 48 insects altogether.

Week 25

Monday: What Doesn't Belong?

A. 6×3.

B. $2 \times \frac{1}{2}$

Tuesday: Vocabulary Tic Tac Toe

Answers vary.

Wednesday: Fraction of the Day

Answers vary. For example:

Word form	Picture form
one and a half	
Plot it on a number line 0, $\frac{1}{2}$, 1, $1\frac{1}{2}$, 2	Is it greater than or less than $\frac{7}{6}$? Yes. How do you know? $\frac{6}{6}$ is equivalent to 1. $\frac{3}{2}$ is 1 and a half so it is more than 1.

Thursday: Number Talk

Answers vary. For example: $\frac{2}{6} + \frac{2}{6}$ or $\frac{1}{6} + \frac{3}{6}$ or $\frac{1}{6} + \frac{1}{6} + \frac{2}{6}$

Friday: Model It

The first piece of wood was 5 times as long as the second piece of wood.

25 feet

5	5	5	5	5

5 feet

5

$25 \div 5$ is 5 . It was 20ft longer.

Week 26

Monday: Open Array Puzzle

457÷8

50	7
400	56

Answer is 57. There is a remainder of 1.

Tuesday: Vocabulary Match

Acute angle: less than 90 degrees
Obtuse angle: greater than 90 degrees and less than 180
Straight line: 180 degrees
Right angle: a 90 degree angle

Wednesday: Guess My Number

$\frac{5}{8}$

Thursday: Open Array Puzzle

100 + 70 + 20 + 14 = 204

Friday: What's the Story?

Answers vary. For example: Mike had 3 marbles. His brother had 3 times as many as he did. How many did they have altogether? They had 12 marbles altogether.

Week 27

Monday: 3 Truths and a Fib

12 is prime is the fib.

Tuesday: Convince Me!

These are not prime numbers because they each have more than 2 factors, meaning 1 and itself. For example, 4 can be made with 2×2. 9 can be made with 3×3 and 15 can be made with 3×5.

Wednesday: Pattern/Skip Counting

4. $\frac{1}{4}, \frac{2}{4}, \frac{3}{4}, \frac{4}{4}, \frac{5}{4}, \frac{6}{4}$
5. $\frac{2}{3}, 1\frac{1}{3}, 2, 2\frac{2}{3}, 3\frac{1}{3}, 4, 4\frac{2}{3}$
6. Answers vary.

Thursday: Number Talk

Answers vary. $\frac{1}{2} + \frac{1}{2} + \frac{1}{2} + \frac{1}{2} = 2$

Friday: Equation Match

The answer is B and the missing number is $\frac{3}{5}$.

Week 28

Monday: Why Is It Not?

It is not $\frac{14}{8}$ because this plus $\frac{6}{8}$ would be $\frac{20}{8}$. It is $\frac{2}{8}$.

Tuesday: 1-Minute Essay

Answers vary. An acute angle is less than 90 degrees.

Wednesday: Guess My Number

$\frac{7}{5}$

Thursday: British Number Talk

Answers vary. For example $3 \times \frac{1}{2} = 1\frac{1}{2}$.

Friday: What's the Question? (3 Read Protocol)

How much butter did she use altogether? $4 \times \frac{2}{4} = \frac{8}{4} = 2$. Answers may vary.

Week 29

Monday: Guess My Number

88.

Tuesday: Vocabulary Bingo

Answers vary.

Wednesday: Bingo

Answers vary.

Thursday: Number Talk

$598 + 573 = 1171$.

Friday: Make Your Own Problem!

Answers vary. For example: Auntie Mary made a pie. She used $\frac{2}{6}$ cup of butter and then she added $\frac{3}{6}$ of a cup more. How much butter did she use altogether?

Week 30

Monday: 2 Arguments

Maria is correct because 42 divided by 7 is 6.

Tuesday: Vocabulary Bingo

Answers vary.

Wednesday: Money Combinations

Answers vary. For example:

Way 1: 16 quarters and a dime and 2 pennies.
Way 2: 4 dollars and 2 nickels and 2 pennies.
Way 3: 8 quarters, $2, 12 pennies.

Thursday: British Number Talk

Answers vary.

Friday: Picture That!

Answers vary. For example: There was a box of donuts. Two-sixths of the donuts were chocolate. What fraction of the donuts were not chocolate?

Week 31

Monday: 3 Truths and a Fib

10 tens and 21 ones.

Tuesday: Vocabulary Brainstorm

Answers vary.

Wednesday: How Many More to

Start at $\frac{1}{4}$ ... Get to $\frac{1}{2}$ ... $\frac{1}{4}$

Start at $\frac{2}{3}$ ... Get to $\frac{9}{6}$ ... $\frac{5}{6}$

Start at $\frac{2}{10}$ ... Get close to $\frac{4}{5}$ ... $\frac{3}{5}$

Start at $\frac{1}{3}$ ... Get to $\frac{9}{12}$ ... $\frac{5}{12}$

Thursday: Number Talk

Answers vary. $\frac{30}{100} + \frac{4}{100} = \frac{34}{100}$

Friday: Fill in the Problem!

Answers vary. For example: Jan ran $\frac{1}{2}$ mile for 3 days in a row. How far did she run? She ran $1\frac{1}{2}$ miles.

Week 32

Monday: Reasoning Matrix

- Jamal: trapezoid.
- Susie: L-shaped hexagon.
- Todd: circle.
- Maribel: rectangle.
- Grace: right angled trapezoid.

Tuesday: Vocabulary Bingo

Answers vary.

Wednesday: Greater Than, Less Than, in Between

Name a fraction less than $\frac{1}{4}$	Name a fraction in between $\frac{2}{3}$ and $\frac{9}{8}$	Name a fraction less than $\frac{2}{3}$
$\frac{1}{5}$	$\frac{3}{4}$	$\frac{1}{2}$
Name a fraction greater than $\frac{9}{8}$	Name a fraction in between $\frac{1}{4}$ and $\frac{2}{3}$	Name a fraction greater than $\frac{2}{3}$
$\frac{4}{2}$	$\frac{2}{4}$	$\frac{6}{6}$

Thursday: British Number Talk

Answers vary. For example: $\frac{3}{10} + \frac{50}{10} = \frac{53}{10}$. I solved it in my head.

Friday: What's the Story? Here's the Model

The bakery made 20 muffins. They put them in 4 boxes. How many muffins did they put in each box? They put 4 muffins in each box.

Week 33

Monday: Draw That!

Answers vary. For example:

1. 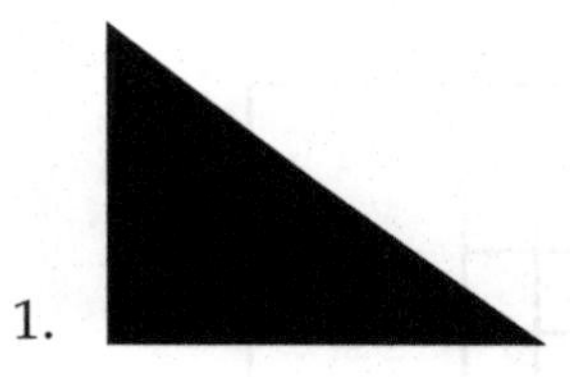

2. 

3. 

Tuesday: Vocabulary Tic Tac Toe

Answers vary.

Wednesday: Model It

Answers vary.

Thursday: What's Missing?

$11 \times 12 = 132$.

Friday: What's the Story? Here's the Graph

Answers vary. For example: The zoo had several lizards. 2 were $\frac{1}{2}$ an inch. 1 was an inch. 4 were $1\frac{1}{8}$ of an inch. 4 were $1\frac{1}{2}$ inches. 3 were $1\frac{3}{4}$ of an inch. 4 were $2\frac{3}{8}$ of an inch. What was the difference between the longest lizard and the shortest lizards?

Week 34

Monday: 2 Arguments

Sara is correct.

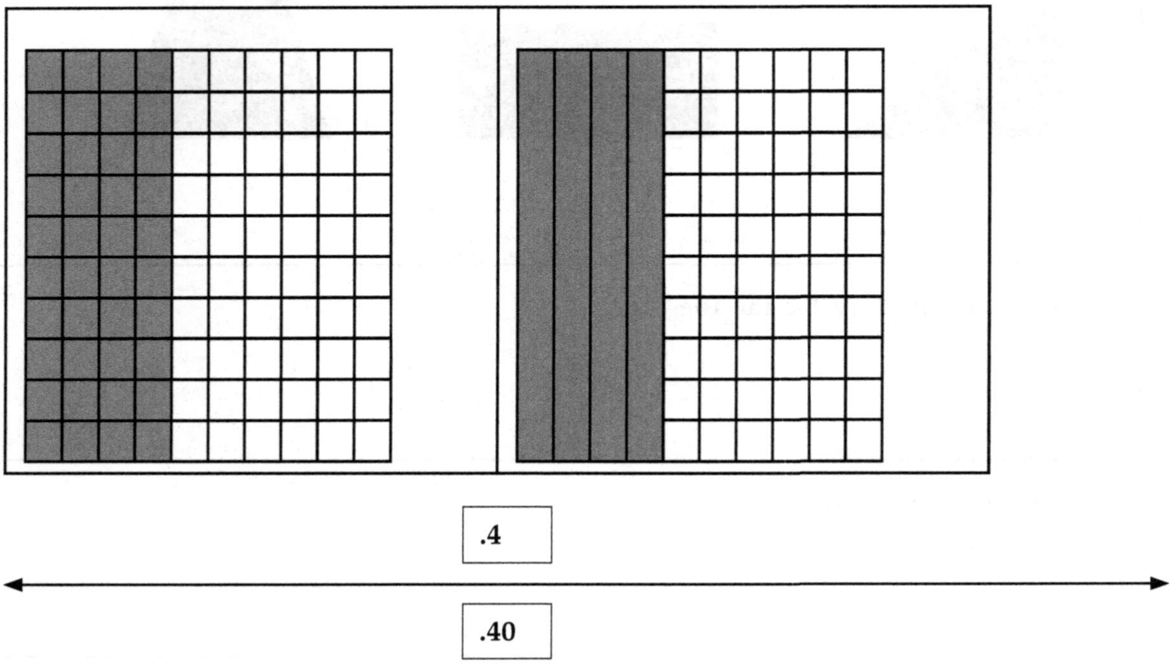

.4

.40

Tuesday: Vocabulary Bingo

Answers vary.

Wednesday: What Doesn't Belong?

A. 8.

B. Trapeziod.

Thursday: Find and Fix the Error

Miguel multiplied and this is a division problem. It should be 50 divided by 5 is 10.

Friday: Model It

Answers vary. For example: Shape A could be a rectangle measuring 1 ft by 12 ft and shape B could be a rectangle measuring 2 ft by 6 ft.

Shape A:	Shape B:
Area 12	Area 12
Perimeter 26	Perimeter 16

Week 35

Monday: Convince Me!

Answers vary. For example: They both equal 8.

Tuesday: Talk and Draw

Answers vary. For example: Parallel lines never meet. Perpendicular lines intersect.

Wednesday: 3 Truths and a Fib

305 is not the same as 3 hundreds and 5 ones.

Thursday: Find and Fix the Error

The correct answer is $\frac{3}{10} + \frac{50}{100} = \frac{80}{100}$. $\frac{53}{100}$ is incorrect because $\frac{3}{10}$ and $\frac{3}{100}$ are different decimals. In this problem we are adding 30 hundredths and 50 hundredths which is $\frac{80}{100}$.

Friday: Regular Word Problem

Answers vary. There could have been 4 tricycles and 4 bicycles.

Week 36

Monday: Subtraction Puzzle

2,192 - 369 = 1,823

Tuesday: Vocabulary Tic Tac Toe

Answers vary.

Wednesday: Number Bond It!

Answers vary. For example: $\frac{7}{8} = \frac{2}{8} + \frac{5}{8}$

Thursday: British Number Talk

Answers vary. For example: $\frac{5}{6} - \frac{3}{6} = \frac{2}{6}$. I can do it using a written strategy.

Friday: Regular Word Problem

Answers vary.

Week 37

Monday: Always, Sometimes, Never

Sometimes. For example, 14 is not a multiple of 4 but 24 is.

Tuesday: Vocabulary Bingo

Answers vary.

Wednesday: Guess My Number

$2\frac{2}{7}$.

Thursday: British Number Talk

Answers vary.

Friday: Fill in the Problem!

Answers vary.

Week 38

Monday: Pattern/Skip Counting

Answers vary. For example: (First task) 35, 42, 40, 47, 45, 52. 50. (Second task) 1,322, 1,329, 1,327, 1,334, 1,332, 1,339, 1,337.

Tuesday: Vocabulary Bingo

Answers vary.

Wednesday: Place Value Puzzle

952 + 830 = 1782

Thursday: British Number Talk

Answers vary.

Friday: Make Your Own Problem!

Answers vary. For example: John had a candy bar. He gave $\frac{1}{5}$ to his brother. He ate $\frac{1}{5}$ How much does he have left? $\frac{3}{5}$

Week 39

Monday: Why Is It Not?

A. It is not 89 because 9×89 does not equal 81. 9×9 equals 81.
B. It is not 18. It is 77 because 77 divided by 11 is 7.

Tuesday: What Doesn't Belong

A. kl
B. grams

Wednesday: Place Value Puzzle

$875 - 123 = 752$

Thursday: British Number Talk

Answers vary.

Friday: Make Your Own Problem!

Larry walked $\frac{2}{10}$ of a mile in the morning and $\frac{4}{10}$ of a mile in the afternoon. If he wanted to walk a full mile, how much more does he need to walk this evening? $\frac{4}{10}$ of a mile.

Week 40

Monday: Missing Number

Answers vary. For example:

- $70 = \frac{700}{10}$
- $100 = 50 \times 2$
- $75 = 25 + 25 + 25$
- $1025 = 1050 - 25$

Tuesday: Vocabulary Tic Tac Toe

Answers vary.

Wednesday: Number of the Day

Answers vary.

Thursday: British Number Talk

Answers vary. For example: $241 \div 6 = 40\frac{1}{6}$

Friday: Make Your Own Problem!

Answers vary. For example: The bakery had 12 boxes of cupcakes with 15 cupcakes in each box. How many cupcakes do they have altogether? 180

Printed in the United States
By Baker & Taylor Publisher Services

Printed in the United States
by Baker & Taylor Publisher Services